THE

ROYAL REGIMENT OF DRAGOONS.

HISTORICAL RECORD

OF THE

FIRST OR THE ROYAL REGIMENT

OF

DRAGOONS

CONTAINING

AN ACCOUNT OF ITS FORMATION
IN THE REIGN OF KING CHARLES THE SECOND,
AND OF ITS SUBSEQUENT SERVICES TO
THE PRESENT TIME

BY

GENERAL DE AINSLIE

COLONEL OF THE REGIMENT

ILLUSTRATED WITH PLATES

WITH

HER MAJESTY'S

GRACIOUS SANCTION

THESE ANNALS OF THE SERVICES OF

THE ROYAL REGIMENT OF DRAGOONS

ARE PUBLISHED,

AND MOST RESPECTFULLY DEDICATED

TO

Our Sovereign Lady Queen Victoria,

IN THIS THE

FIFTIETH YEAR OF HER REIGN.

GENERAL ORDERS.

HORSE GUARDS, *1st January*, 1836.

HIS MAJESTY has been pleased to command that, with a view of doing the fullest justice to regiments, as well as to individuals, who have distinguished themselves by their bravery in action with the enemy, an account of the services of every regiment in the British Army shall be published under the superintendence and direction of the Adjutant-General; and that this account shall contain the following particulars, viz :—

The period and circumstances of the original formation of the regiment. The stations at which it has from time to time been employed. The battles, sieges, and other military operations, in which it has been engaged; particularly specifying any achievements it may have performed, and the colours, trophies, etc., it may have captured from the enemy. The names of the Officers, and the number of Non-commissioned Officers and Privates killed or wounded by the enemy; specifying the place and date of the Action. The names of those

Officers who, in consideration of their gallant services and meritorious conduct in engagements with the enemy, have been distinguished with titles, medals, or other marks of His Majesty's gracious favour.

The names of all such Officers, Non-commissioned Officers, and Privates as may have specially signalized themselves in Action, and

The badges and devices which the regiment may have been permitted to wear, and the causes on account of which such badges or devices, or any other marks of distinction, have been granted.

By command of the Right Honourable

GENERAL LORD HILL,

Commander-in-Chief.

JOHN MACDONALD,

Adjutant-General.

INTRODUCTION.

"SPECTEMUR AGENDO" is a proud motto, and a bold: it is one nevertheless which I believe all who may read the following pages, in which it is proposed to relate the long, historic, and eminent services of one of the most distinguished regiments in the British army, will admit may be borne by "The First" or "The Royal Regiment of Dragoons" with equal pride in the Past and confidence in the Future.

Since the origin of the corps in 1661 to the present day, "The Royal Dragoons" have during this long period invariably upheld the honour of their country and the character of the service to which they belong upon many trying and memorable occasions, and have exhibited in a striking degree the great military virtues of Loyalty to their Sovereign, steady unswerving Discipline, Efficiency, and that determined Valour which is in truth the rarely failing characteristic of the British soldier.

In their annals will be found not only the records of their military and Continental achievements, but also of those various and often difficult services in which they have been employed at home, in all which circumstances they have discharged their important duties with temper, forbearance, and firmness ; and it cannot fail also to be noticed how frequently, in times of trouble and danger, the Royal Regiment of Dragoons has been one of the first to be brought to the immediate vicinity of the Sovereign, as a protection to his person and a support to the Government.

From 1664 to 1680 the capabilities and value of the Royal Dragoons were first tried in their severe conflicts with the Moors in Africa. Later on, in 1688, at Sedgemoor, in routing the insurgent bands of the Duke of Monmouth ; in forcing the passage of the Boyne under King William III. in 1690 ; and in subsequent detached operations in Ireland in 1691. From 1694 to 1697 they served with credit against the armies of Louis XIV. in the Netherlands, and in 1702-3 under the great Duke of Marlborough on the frontiers of Holland. They made the campaigns in Spain of 1705-6 under the Earl of Peterborough, and shared in the glories of Almanara on the 27th of July, and Saragossa on the 20th of August, 1710.

During the disturbances in Scotland in the years 1718 and 1719 the regiment was actively employed, and in the war in Germany in 1742-45 they highly distinguished themselves at the battle of Dettingen under the eyes of King George II., taking there the standard of the French "Mousquetaires Noirs." They were also engaged at Fontenoy. From 1760 to 1763 they were again in

Germany, and behaved with equal gallantry at Warbourg on the 31st of July, 1760.

We find the Royal Dragoons once more in Flanders in 1794 with the army under H.R.H. the Duke of York, and afterwards, throughout the long and arduous contest in the Peninsula with the legions of Napoleon, they acquired additional reputation from 1810 to 1814, during which years they were constantly, and often particularly engaged, and notably in the month of September, 1810, in covering the retreat of the allied army from Busaco to the celebrated lines of Torres Vedras; at Fuentes d'Onor, on the 8th of May, 1811, and at the brilliant affair at Gallegos on the 6th of June ensuing; throughout the constant and harassing service in Spanish Estremadura in 1812; in a spirited affair on the 26th of May, 1813, near Salamanca; and throughout the operations consequent upon the battles of Vittoria and Toulouse, until the final close of the Peninsular war in 1814.

The conduct of the Royal Regiment of Dragoons during the short but transcendent campaign of 1815 in the Netherlands, ending in the closing triumphs of Waterloo, in which they captured one of the two "Eagles" taken that day, is written in characters which can perish but with the world itself, and the part lately taken by the small detachment of the corps employed in the operations in Egypt in 1884-85 is worthy of its reputation.

In concluding these preliminary observations, and in

leaving the further continuance of the story of this ancient and illustrious regiment to future, and I hope abler, hands, it may be permitted me, I trust, to express my earnest pride and gratification to have so long been identified with the noble corps at whose head, by Her Majesty's gracious favour, I have for many years been placed.

CHARLES P. DE AINSLIE, General.

Colonel of the Royal Regiment of Dragoons.

CONTENTS.

LIST OF ILLUSTRATIONS.

Designed and Modelled by Carrington and Co.

THE FIGHT FOR THE STANDARD.

"ROYAL SCOTS GREYS" WATERLOO.

"SPECTEMUR AGENDO."

THE FIRST,

OR

THE ROYAL REGIMENT OF DRAGOONS,

BEARS ON ITS GUIDON THE REGIMENTAL BADGES OF

"THE CREST OF ENGLAND WITHIN THE GARTER,"

"AN EAGLE,"

WITH THE HONORARY INSCRIPTIONS—

"DETTINGEN." "PENINSULA."

"WATERLOO." "BALACLAVA."

"SEVASTOPOL."

1660

Vincent Brooks, Day & Son, Lith

1742

m Brooks, Day & Son, Lith

1751

1809

1815

1825

1839

1886

THE

ROYAL REGIMENT OF DRAGOONS.

CHAPTER I.

FORMATION OF THE REGIMENT.

In compliance, therefore, with his Majesty's order of 1st January, 1836, the Records of the several regiments of the Army were undertaken by Mr. Richard Cannon, principal clerk of the Adjutant-General's Office, under the direction of the Adjutant-General; and it must be admitted that in their compilation and arrangement great intelligence and general accuracy are conspicuous; and from them, it will be seen in the following narrative, large extracts have been made. These Records, however, are frequently meagre and insufficient; and moreover since their publication a very long period has gone by, during which a variety of events have occurred materially affecting the circumstances of the Army; it is hoped that in the history now presented of the services and fortunes of a corps so ancient and of such historical interest as the Royal Regiment of Dragoons, the existing deficiencies may be found in a great degree made up, and the story carried on with some success to the present date.

It is not intended to enter too deeply into the original nature of the regular forces, or Standing Army, of Great Britain; but before arriving at the particular account of the Royal Dragoons, it is necessary briefly to introduce the subject by recapitulating that on the restoration of the Monarchy, in 1660, one of the earliest cares of the ministers of King Charles II. was the formation and consolidation of a Standing Army, upon which the Government and the country might in all future times with confidence rely.

The veteran, well disciplined soldiers, whether Cavaliers or Parliamentarians, who had fought through the struggles of the Civil War, and under the Protectorate of Cromwell, supplied material of the best description, which only required to be organised and placed upon a permanent footing; since which early period of their history the conduct of the British troops, their valour and efficiency, have been notorious, it may be truly said, in every quarter of the globe.

Confining ourselves, however, to the Cavalry, it consisted originally of the Life Guards, of Horse, and subsequently of some corps of Dragoons, which latter were troops supposed to combine the advantages of serving either on foot or horseback, with which object they were equipped rather as infantry, mounted upon a smaller description of horses, and placed upon a lower rate of pay than the regiments of Horse; but it was found that Dragoons thus constituted fell rapidly into disrepute, and consequently ere long they were put in all respects upon an equality with the rest of the Cavalry.

The formation of the Life Guards dates from the Restoration, when it appears that on the 2nd of April 1661, the King's Life Guard of 120 Noblemen and Gentlemen was formed in Scotland, under the command of Lord Newburgh, being the nucleus of the Regiment of the famous John Graham of Claverhouse, Viscount Dundee, which, after the union of the two kingdoms in 1707, was removed to London, and is now represented by the 2nd troop of the 1st Life Guards. The Royal Horse Guards, or The Blues, had been raised by royal warrant of 16th of February, 1661, and were then styled the Royal Regiment of Horse.

In 1666 eighty Cavalier Gentlemen who had adopted the profession of arms, and had followed the fortunes of King Charles I. during the Civil Wars, were embodied into a Guard for the protection of the royal person, under the command of Lord Everard, afterwards Earl of Macclesfield. The arms both of defence and offence of the Household Cavalry, and of the regiments of Horse, are thus laid down in the Regulations of King Charles II., dated 8th of May, 1663 :—

"Each horseman to have for his defensive arms, back breast and pott; and for his offensive arms a sword and a case of pistols, the barrels thereof are not to be under fourteen inches in length; and each trooper of our Guards to have a carbine besides the aforesaid arms."

The composition and appearance of the Household Cavalry at this time are thus described by Lord Macaulay in his *History of England* (vol. i. chap. iii.) :—

"The Life Guards, who now form two regiments, were then distributed into three troops, each of which consisted of 200 Carbiniers, exclusive of officers. This corps, to which the safety of the King and royal family was confided, had a very peculiar character. Even the privates were designated as 'Gentlemen of the Guard.' Many of them were of good families, and had held commissions during the Civil War. Their pay was far higher than that of the most favoured regiment of our time, and would in that age have been thought a respectable provision for the son of a country squire. Their fine horses, their rich housings, their cuirasses, and their buff coats adorned with ribbands, velvet and gold lace, made a splendid appearance in St. James's Park. A small body of grenadier dragoons, who came from a lower class and received lower pay, was attached to each troop. Another body of Household Cavalry, distinguished by blue coats and cloaks, and still called 'The Blues,' was generally quartered in the neighbourhood of the capital."

The old and original regiments of Horse formed a highly efficient and respectable portion of the Army; and the records of these troops, embracing an eventful period of 180 years, show that on all occasions, both at home and abroad, they have maintained a high character for discipline and good conduct. Their ranks were composed of men of some property, generally sons of substantial yeomen, who provided their own horses, and they were placed on a rate of pay sufficient to give them a respectable station in society.

Upon the incorporation with the Household Cavalry of the Royal Regiment of Horse, as the Royal Horse Guards, the 2nd or Queen's regiment of Horse became the first of the three regiments of Horse on the English establishment, the remaining four being on the Irish. And here it is to be explained with regard to these two establishments that by a permanent statute of King William III. the so-called "Standing Army of Ireland" was constituted, which in the reign of George III. was further increased to 15,234 men, and was in force at the union of the kingdoms in 1801. The regiments composing this establishment varied from time to time as they proceeded to or were removed from Ireland, except in cases where regiments on the Irish establishment went on active service. The separate English and Irish establishments ceased at the Union.

In the year 1746 King George II. reduced the three old regiments of Horse on the English establishment to the quality and pay of Dragoons, His Majesty at the same time conferring upon them the respective titles of the 1st or King's Dragoon Guards; the 2nd or Queen's Dragoon Guards, or Queen's Bays; and the 3rd or Prince of Wales's Dragoon Guards. In like manner King George III. reduced the four regiments of the Irish establishment, which then became, from the 1st of April, 1768, the 4th or Royal Irish Dragoon Guards; 5th or Princess Charlotte of Wales's Dragoon Guards, 6th Dragoon Guards, or the Carbineers, and 7th or the Princess Royal's Dragoon Guards. There are at present but three regiments of Horse properly so called in the Army, viz.: the two regiments of Life Guards and the Royal Horse Guards, whose Non-commissioned Officers are still styled Corporals of Horse.

The first regiment of Dragoons seems to have been raised on the breaking out of the war with Holland, in the spring of 1672, agreeably to a royal warrant of the 2nd of April of that year, of which the following is an extract:—

"CHARLES R.

"Our will and pleasure is, that a regiment of Dragoons, which we have established and ordered to be raised in twelve troops of four score in each, besides officers, who are to be under the command of our most deare and most entirely-beloved cousin, Prince Rupert,—shall be armed out of our stores remaining within our office of the Ordinance as followeth:—that is to say, three corporalls, two sergeants, the gentlemen at armes, and twelve souldiers of each of the twelve Troopes, are to have, and to carry each of them one halbard and one case of pistolls with holsters; and the rest of the souldiers of the twelve Troopes aforesaid are to have and to carry each of them one matchlocke musquet, with a collar of bandaliers and also to have and to carry one bayonett or great knife. That each lieutenant have and carry one partizan; and that two drums be delivered out for each Troope of the said Regiment."

This appears to have been the first introduction of bayonets into the British Army. Carbines do not seem to have been supplied to the regiments of Horse until 1678. This regiment was disbanded after the peace in 1674.

In the first year of the reign of King James II. several regiments of Horse and Dragoons were embodied; and a regulation of the 21st of February, 1687, thus prescribes the arms of the Dragoons:—

"The Dragoons to have snaphause musquets strapt with bright barrels of three foote eight inches long, cartouch boxes, bayonetts, grenade pouches, bucketts and hammer hatchetts."

However, as already observed, the service of Dragoons thus constituted soon became unpopular, and was in no degree found to answer the purposes intended, so that in all respects they were ere long assimilated to the rest of the Cavalry. This became the more desirable, as the suppression generally of the cuirass—restored to the Household regiments only in 1821—and the improvements in mounting and equipment had, in fact, placed the service pretty much on the same footing. The introduction of regiments of Light Dragoons originated in 1748 by the formation of the Duke of Kingston's regiment of Light Horse. In 1755 King George II. added Light Troops to certain regiments; and in 1759 were raised special corps of Light Dragoons, of which the 15th was the first, being also one of four regiments of Hussars so constituted in 1803, when they received the title of the 15th or King's Hussars. The lance was introduced in 1817; the first corps so armed being the 23rd Light Dragoons. In 1862 the title of the regiments of Light Dragoons was converted into that of Hussars, although the peculiar dress and equipment which had distinguished these corps had for some time been abolished.

In the year 1664, King Charles II. having contracted an alliance with Donna Catherina, Infanta of Portugal, and receiving as her marriage portion a sum of money equal to £300,000, together with the island of Bombay in the East Indies and the city of Tangiers in Africa, this last acquisition, with its important fortress, its harbour and local advantages, appeared to open out a new field for commercial enterprise, to be followed, it was expected, by the acquirement of extensive possessions in that country, and in consequence a garrison of four regiments of Foot and a troop of Horse was appointed to that place, of which the Earl of Peterborough was constituted Captain-General, Chief Governor, and Admiral.

Three of the regiments of Foot, commanded respectively by Sir Robert Harley and Colonels Fitzgerald and O'Farell, were taken from the garrison of Dunkirk; the other regiment, now the 2nd or Queen's Royal, and the troop of Horse, the nucleus, as will be subsequently seen, of the Royal Regiment of Dragoons, had been raised in England by the Earl of Peterborough in the autumn of 1661, and were mustered, the former on Putney Heath, and the latter in St. George's Fields, Southwark, in October that year.

The troop of Horse consisted of three officers, one quartermaster, four corporals, and 100 privates; the ranks were completed with veterans of the Civil War, who were armed with cuirasses, iron headpieces called potts, long swords, and a pair of large pistols to which a short carbine was afterwards added. They were mounted upon long-tailed horses of superior weight and power; wore high boots reaching to the middle of the thigh, and scarlet vests. The officers wore hats decorated with a profusion of feathers, and both officers and men ornamented their horses' heads and tails with large bunches of ribbons.

The officers of this troop were the Earl of Peterborough, Captain and Colonel; Robert Leech, Captain-Lieutenant; James Mordaunt, Cornet.

The appearance and equipment of the officers and men were much commended in the publications of that period. They embarked in the middle of December, 1661 ; and in a letter to the Earl of Peterborough, dated the 20th of the month, the King observes :—

"I desire you to lett those honest men knowe who are along with you y^{t} they shall allways be in my particular care and protection as persons y^{t} venture themselves in my service; and so, wishing you a good voyage,

"I remain, &c.,

"CHARLES R."

The troops arrived at Tangiers in January, 1662, and war commencing soon afterwards between the Britis in this part of Africa and the Moors, frequent encounters took place between the garrison and the barbarians, to the decided advantage of the former, and in which the English horsemen became celebrated for gallant conduct.

In 1663 the veteran Earl of Teviot, who had been appointed Governor of Tangiers in succession to the Earl of Peterborough, occasionally penetrated into the adjacent country at the head of a party of Horse, who performed many brilliant exploits on the neighbouring plains and among the rocks and woods, where they frequently surprised lurking bodies of the Moors and made captures of cattle and other spoil. These

Africans, however, were clever horsemen, and fought with lances, swords, and short fusils.

In February, 1664, a Moorish army, commanded by Gaylan the usurper of Fez, appeared before Tangiers with the object of laying siege to the fortress. On the 1st of March the Earl of Teviot observing a body of the enemy, with a splendid scarlet standard, on an eminence near the city, ordered the troop of Horse to make a sally and bring in the standard, which command being promptly obeyed, the brave troopers, led by Captain Witham, issued from the city, traversed the intervening space with signal intrepidity, and, having routed the Moors, they returned in triumph with the standard, which they hoisted on one of the towers of the fortress, to the surprise and chagrin of the Moorish chiefs, who from a distance with the main body of their army had witnessed this feat of arms.

On the 13th of March the Cuirassiers had a smart affair with some of the enemy's best Cavalry; and on the 27th the Earl of Teviot in person led them against a horde of Lancers and Foot, who were lying in ambush, when the barbarians were routed and pursued among the rocks and broken ground with great slaughter. On the 4th of May, however, the English met with a severe repulse, when the Governor, deceived by a false report, advanced too far into the interior, and, being surprised by a numerous band of Moors, a fearful massacre ensued, and the gallant Earl of Teviot was numbered among the slain.

Frequent affairs happened during the subsequent years between the English and the Moors, in which desultory warfare the troop of Horse continued to maintain its high character. Hostilities were occasionally suspended and renewed after short intervals of peace, and during seventeen years the garrison of Tangiers resisted with success every attempt made upon the city.

In 1679 a numerous army appeared before Tangiers, and destroyed the forts constructed at a distance from the city, after which they withdrew, but reappeared in the spring of 1680, in increased numbers and with swarms of clever horsemen on light and swift horses, who, hovering round the walls, confined the Christians within narrow limits. King Charles II. despatched a battalion of the Foot Guards and sixteen companies of Dumbartons, now the 1st or Royal Scots Regiment, to reinforce the garrison, and issued commissions for raising in England a regiment of Foot, now the 4th or King's Own, and six troops of Horse, while at the same time arrangements were made for procuring the services of three troops of Spanish Cavalry.

The six troops of English Horse were raised respectively by Major-General the Earl of Ossory; Colonel Sir John Lanier; Captains Robert Pulteney, John Coy, Charles Nedby, and Thomas Langston. The three last-named officers having been Captains in the Duke of Monmouth's regiment of Horse—which had been disbanded only a few months before—their troops were speedily completed with disciplined soldiers who also had served in that regiment, and the demand for Cavalry at Tangiers being urgent they were at once supplied with horses and equipment from the Life Guards, and arrived at Tangiers in the early part of September, 1680, at the same time as the three troops of Spanish Cavalry arrived there from Gibraltar.

The Cavalry at Tangiers now consisted of seven troops of efficient Cuirassiers, who were engaged on the 12th of September, when the Moorish Horse were driven from under the walls and several outworks of the fortifications were recovered. Another sally was on the 20th, and on the following day the Cuirassiers had a smart skirmish with the Moorish Lancers and had eight men killed and twenty wounded. An attack on the enemy's lines was made on the 24th, when the Governor, Sir Palmes Fairborne, was mortally wounded.

On the 27th of September the garrison, amounting to about 4,000 men, issued from the fortress and attacked the army of the Moors, estimated at 18,000 men, in their entrenched camp with signal audacity. So eager was the Cavalry to engage that a dispute actually arose between the English and Spanish Horse, each claiming the honour of making the first charge, when the matter being referred to the Lieutenant-Governor, Colonel Sackville, he gave the Spaniards the precedence because they fought as "auxiliaries." The Moors having a great superiority of numbers stood their ground for some time with much resolution, and the thunder of artillery, the roll of musketry, the clash of arms, the loud shouts of the British and the wild cries of the Africans produced an awful scene of carnage and confusion. The English and Spanish Horse stood in column of troops until the first entrenchment was carried and a space levelled for the passage of the Cavalry, when they dashed through the opening and rushed at full speed upon the dark masses of the defenders, who were broken, trampled down, and pursued with dreadful slaughter, while the musketeers, pikemen and grenadiers followed with loud shouts as the dismayed Africans fell beneath the sabres of the English and Spanish troopers. Many of the Moors faced about and confronted their pursuers; numerous single combats took place, and the vicinity of the camp was covered with slain. Captain Nedby's troop of English Horse particularly distinguished itself, and captured a standard of curious workmanship. The Spaniards also captured a colour, Dumbarton's "Scots" another, and a fourth was taken by a battalion of marines and seamen from the fleet.

The Moorish legions having been driven from before the city with severe loss, this victory was followed by a treaty of peace, when the troops of Horse raised by the Earl of Ossory, Sir John Lanier, and Captain Pulteney, not having left England were disbanded.

The improved military system introduced among the Moors by European renegades rendering it now necessary to employ at Tangiers a much stronger garrison than hitherto, the question was brought before Parliament, but no grant of money being voted, it was decided by the Government to destroy the works and withdraw the troops.

At this period the attention of King Charles II. was particularly directed to the improvement of his army, and resolving to retain the services of the Tangiers Horse, His Majesty commissioned Colonel John Churchill to raise a troop of Dragoons at St. Albans and its vicinity; and Viscount Cornbury, son of the Earl of Clarendon, to raise another at Hertford; and His Majesty constituted these two, with the four troops of Tangiers Horse, a regiment to which he gave the distinguished title of "The King's Own Royal Regiment of Dragoons"; the words "King's Own" were, however, soon afterwards discontinued, and the regiment was styled "The Royal Regiment of Dragoons." In 1672 a corps had been raised bearing this title, but it was disbanded after the Peace of Nimeguen in October of that year. The Colonelcy of the new regiment was conferred upon Colonel Churchill, now advanced to the dignity of Baron Churchill of Eyemouth, by commission, dated 19th of November, 1683, and the Lieut.-Colonelcy at the same time upon Viscount Cornbury.

"CHARLES R.

"Our will and pleasure is that as soon as the troop of our Royal Regiment of Dragoons, whereof Charles Nedby, Esq., is captain, shall arrive from our garrisons at Tangiers, you cause the same forthwith to march to the town of Ware, in our county of Hertford, where they are to remain until further orders; and the officers of the said troops are to take care that the soldiers duly pay their intended quarters.

"Given at our Court at Whitehall, this 18th day of February, 1623.

"By His Majesty's command,

"WILLIAM BLATHWAYTE."

A similar order was given for Captain Thomas Langston's troop to quarter at Hoddesdon, Captain John Coy's at Hampstead, and Captain Alexander Mackenzie's (the troops raised in 1661) at Watford and Bushey.—*War Office Records.*

The establishment was fixed, by a warrant bearing date the 18th of January, 1684, from which the following is an extract:—

"CHARLES R.

"Charles the Second, by the Grace of God King of England, Scotland, France and Ireland, Defender of the Faith, &c.

"Our will and pleasure is that this establishment of our Guards, Cuirassiers, and land forces within our Kingdom of England, Dominion of Wales, and Town of Berwick-upon-Tweed and the Islands thereunto belonging, and of all other offices and charges therein expressed, do commence on the 1st day of January, 1683—4, in the thirty-fifthe year of our reign.

"His Majesty's own Royal Regiment of Dragoons."—10*th page*, *Records C.*

HIS MAJESTY'S OWN ROYAL REGIMENT OF DRAGOONS.

Staff Officers.	Per Diem. £	s.	d.
Colonel, *as Colonel*, xiis, and iij horses iijs	0	15	0
Lieutenant-Colonel, *as Lieut.-Colonel*, vijs, and ij horses ijs	0	9	0
Major, as Major vs, and j horse js	0	6	0
Chaplaine	0	6	8
Chirurgeon ivs, and j horse to carry his chest, ijs	0	6	0
Adjutant ivs, and for his horse js	0	5	0
Quarter-Master and Marshal in one person ivs, his horse js	0	5	0
Gunsmith ivs, and his servant is	0	5	0
	2	17	8
The Colonel's Troop.			
The Colonel, *as Captaine*, viiis, and iij horses iijs	0	11	0
Lieutenant ivs, and ij horses ijs	0	6	0
Cornett iijs, and ij horses ijs	0	5	0
Quarter-Master, for himself and horse	0	4	0
Two Serjeants each js vid, and ijs for horses	0	5	0
Three Corporals each js, and iijs for horses	0	6	0
Two Drummers each js, and ijs for horses	0	4	0
Two Hautboys each is, and ijs for horses	0	4	0
Fifty Soldiers each is vid for man and horse	3	15	0
	6	0	0
Five Troops more at the same rate	30	0	0
The Major to have no Troop, but instead thereof the pay of a Captain xis, in lieu of servants iiis	0	14	0
Total	39	11	8
Total per Annum . . . . £14,447 18*s.* 4*d.*			

The four troops from Tangiers arrived in England in February, 1684, and having returned their cuirasses into store, the whole were equipped as dragoons, and the following arms and appointments were issued to the regiment from the Tower of London, viz.,

316 Muskets and bayonets.
12 Halberds.
12 Partisans.
12 Drums.
316 Cartouch boxes and belts.
318 Waist belts and bayonet frogs.
358 Saddles and bridles.
388 Sets of holster caps and housings.
—*War Office Records.*

The uniform of the regiment was scarlet lined with blue. The men wore hats bound with silver lace, and ornamented with ribbons, having a metal head-piece fastened inside the crown; also high boots. Their horse furniture was of scarlet cloth trimmed with blue, with the King's cipher embroidered in yellow on the housings and holster caps. The drummers and hautboys were clothed in splendid uniforms, which, according to the *War Office Records,* cost upwards of £10 per suit, and each troop was furnished with a crimson standard or guidon, having the following devices embroidered thereon, viz.:—

On the standard of the colonel's troop: the King's cipher and crown; the lieutenant-colonel's troop the rays of the sun, proper, crowned, issuing out of a cloud, proper—a badge used by the Black Prince. The first troop: the top of a beacon crowned, or, with flames of fire, proper—a badge of Henry V.

The second troop: three ostrich feathers, crowned, argent—a badge of Henry VI.

The third troop: a rose and pomegranate impaled, leaves and stalk vert—a badge of Henry VII.

The fourth troop: a phœnix in flames, proper—a badge of Queen Elizabeth.[1]

The following officers were at this period holding commissions in the regiment:—

TROOPS.	CAPTAINS.	LIEUTENANTS.	CORNETS
Colonels . .	Lord Churchill.	Thos. Hussey.	Wm. Hussey.
Lieut.-Cols. .	Visc. Cornbury.	Charles Ward.	Piercy Roche.
1st Troop . .	Alex. Mackenzie.	H. Wyndham.	John Cole.
2nd ,,	Chas. Nedby.	John Williams.	George Clifford.
3rd ,,	John Coy.	Charles La Rue.	William Stamford.
4th ,,	Thomas Langston.	F. Langston.	Thomas Pownel.

Hugh Sutherland	Major.
Thomas Crawley	Adjutant.
Henry Hawker	Quarter-Master and Marshal.
Theobald Churchill	Chaplain.
Peregrine Yewel	Chirurgeon.

Lieutenant Hugh Wyndham became afterwards colonel of the 7th Horse, the present 6th Dragoon Guards, the Carbineers.

Lieutenant Francis Langston was subsequently colonel of the 5th Horse, 4th R. I. Dragoon Guards.

The Royal Regiment of Dragoons being thus formed, and composed as it was generally, of men of

[1] Nathan Brook's *Complete List: Military.* London, 1684.

approved valour and military experience, appears to have advanced at once into royal favour, and as soon as it was completely organised it went into quarters in the borough of Southwark. On the first day of October it was reviewed with several other corps by King Charles II., by the Queen, the Duke of York, and many distinguished personages, on Putney Heath, and Lord Macaulay has the following notice of the regiment in his *History of England* (vol. i. chap. iii.) :—

"Near the capital lay also the corps which is now designated as the 'First' Regiment of Dragoons, but which was then the only regiment of dragoons on the English establishment. It had recently been formed out of the cavalry which had returned from Tangiers."

On the 13th of October the Royal Dragoons marched into quarters at Newbury, Abingdon and Hungerford, and shortly afterwards the following order was issued by his Majesty's command :—

"CHARLES R.

"For the preventing of all disputes that might arise concerning the rank of our Royal Regiment of Dragoons, or of any other regiment of dragoons that shall be employed in the service, we have thought fit hereby to declare our pleasure—

"That our Royal Regiment of Dragoons, and all the regiments of dragoons which may be employed in our service, shall have precedence both as horse and foot, as well in garrison as in the field, and in all councils of war and other military measures; and the colonels and officers of the said regiments of dragoons shall command as officers of horse and foot according to the nature of the place where they shall be; that is to say, that in the field the said regiments shall take place as regiments of horse, and the officers shall command and do duty as officers of horse according to the dates of their commissions, and that in garrison they shall command as foot officers, and their regiments take place amongst the foot according to their respective seniorities from the time they were raised.

"Given at the Court at Whitehall, the 30th day of October, in the thirty-sixth year of our reign, 1684.

"By his Majesty's command,

"SUNDERLAND."

The decease of King Charles II. took place on the 6th of February, 1688, and that same evening his successor James II. gave orders for the Royal Regiment of Dragoons to be brought to the immediate vicinity of the capital.

Previous to the ceremonial of the coronation of James and his Queen, which was celebrated with extraordinary magnificence on the 23rd of April, the regiment received new guidons, and the drummers and hautboys new uniforms;[1] but the agitated condition of the country gave early indication of approaching troubles, and at the end of the month two troops of the regiment were despatched

[1] *War Office Records.*

to Carlisle, where they arrived on the 10th of May, and were placed under the orders of the governor, Sir Christopher Musgrave, for the purpose of assisting in the seizure of "divers outlawed and seditious persons who, for the avoiding of justice, have fled from Scotland into the county of Cumberland and parts adjacent,"[1] where several persons were apprehended.

In the middle of that month an insurrection broke out in Scotland, headed by Archibald, third Earl of Argyll, who, being taken on the 18th of June, was beheaded on the 30th following, while in the meantime James, Duke of Monmouth, had raised the standard of revolt in the west of England, at Lyme, in Dorsetshire, and proclaimed himself King. The establishment of the Royal Dragoons was immediately augmented to sixty men per troop; an independent troop of dragoons, raised by Colonel Strother in 1663, was incorporated in the regiment, and five troops were raised in the neighbourhood of London by Richard Leveson, John Williams, Edward Lea, Francis Russel and Thomas Hussey, and added to the corps, whose numbers were thus increased to twelve troops, amounting to about 900 officers and men.

A squadron of the regiment with some other forces was despatched under Brigadier-General Lord Churchill against the rebels in the west; and on the 19th of June another squadron marched for the same destination under the orders of Lieutenant-General the Earl of Feversham, who was appointed to the command-in-chief

[1] *War Office Records.*

of the King's army. The royal forces having united, the four troops of dragoons were placed under Lieutenant-Colonel the Viscount Cornbury, when the whole marched against the rebels.

After some marching and skirmishing the Duke of Monmouth took post at Bridgewater, in Somersetshire, while the Earl of Feversham, having sent a troop of the Royal Dragoons under Captain Coy to Lamport to secure that pass, advanced with the royal army to Weston, about three miles from Bridgewater, where he arrived on the 5th of July. Quartering the cavalry in the village, his lordship encamped his infantry on a plain, having in front the wild tract of Sedgemoor, between Weston and Bridgewater. He sent a patrol of the Life Guards in the direction of Bristol, and posted a picket of fifty men of the Royal Dragoons with a squadron of the Blues, supported by 100 men of the Royal Regiment of Foot, on the moor in front of the camp. A guard also of the Royal Dragoons was posted over the artillery, which consisted of sixteen pieces, and was drawn up on the high road from Weston to Bridgewater.

At two o'clock on Monday morning the Duke of Monmouth at the head of his infantry marched out of Bridgewater with the view of surprising the royal forces in their position, but unexpected obstacles delayed his march; a random shot alarmed the picket in advance, who, after exchanging a few shots with the rebels, fell back upon the camp and formed upon the right of the infantry: at the same time the remainder of the Royal Dragoons, aroused in their quarters in the village of Weston, turned out in the dark and formed on the left of the Foot. The rebels commenced the attack with loud shouts; the contest became general along the whole line, and Sedgemoor sparkled with fire. The duke's Horse, commanded by Lord Grey, were soon dispersed and fled, but the Foot stood firm and fought with great resolution. Day beginning to break, the King's Foot advanced to the charge, while the Royal Dragoons and the cavalry, falling upon the flanks of the rebels, their whole line gave way and fled, being pursued across the moor and adjoining cornfields with great slaughter. Two troops of the Royals continued the pursuit as far as Bridgewater, where they were ordered to halt by the Earl of Feversham.

In the meantime Captain Russel's troop of the regiment had been attached to three Scots regiments of Foot, recently arrived from Holland under the command of Major-General Mackay, and ordered to join the army in the west, but on the news of the battle of Sedgemoor these forces were halted at Bagshot.

The troop of the Royal Dragoons was subsequently dispersed in small parties into the adjoining counties to seize suspected persons. The Scots regiments went to Hounslow, whence, after encamping for a short time on the heath there, they re-embarked for Holland.

One troop of the Royal Dragoons was ordered to

Winchester to escort the Duke of Monmouth and other prisoners to London, where on arrival it was quartered in Southwark, and was on duty on the 15th of July, when the duke was beheaded on Tower Hill. Two other troops were sent to Salisbury to mount guard over the prisoners there, and afterwards to attend Judge Jeffries during the trial and execution of the captured rebels, in the course of which painful service the soldiers witnessed the many acts of barbarity perpetrated by the remorseless judge who sacrificed 320 lives during these "Bloody Assizes," as they are denominated by historians.

After the suppression of this rebellion the establishment of the Royal Regiment of Dragoons was reduced to eight troops of forty men each, and the supernumerary troops, together with one independent troop, were embodied into a regiment styled "The Queen Consort's Regiment of Dragoons," of which Charles, Duke of Somerset, was appointed colonel, and which corps is the present 3rd King's Own Hussars.

On the 1st of August this year Lord Churchill was appointed colonel of the 3rd troop of Life Guards, when the colonelcy of the Royal Dragoons was conferred upon Lieutenant-Colonel Viscount Cornbury. The two troops of the regiment returning from Carlisle, the whole concentrated in London in October, and marched subsequently into quarters in Devonshire.

King James II., being a Roman Catholic, now proceeded to the adoption of measures calculated for the subversion of the Protestant Church, and with the view of overawing his subjects he largely increased the strength of the army, and caused considerable bodies of troops to encamp on Hounslow Heath, where his Majesty frequently attended in person to witness their exercise. The Royal Regiment of Dragoons made part of the force at these camps in the summers of 1686, 1687, and 1688.

The strained and ill feeling which had for some time existed between the sovereign and the nation at length rose at this period to such a height that several influential personages in the country, determined to resist the encroachments of James and the Papists by whom he was surrounded, addressed an invitation to William, Prince of Orange, in compliance with which that Prince, at the head of a Dutch force, landed at Torbay, on the coast of Devonshire, on the 5th of November, 1688. To oppose this aggression the army of King James was ordered to assemble at Salisbury,[1] whither Lord Cornbury proceeded with the Royal Dragoons, but being himself a zealous Protestant, his lordship entered warmly into the anti-Papist movement, and finding at Salisbury the Blues and the Eighth Horse, he determined, in concert with Lieutenant-Colonel Langston of the 8th, and several officers of the Blues, to take these three regiments over to the Prince of Orange in the following summer.

[1] *War Office Records.*

On the night of the 10th of November, upon the arrival of the post, at twelve o'clock, Lieutenant-Colonel Langston, in presence of the officers, opened the letter bag, when the orders, apparently from the Secretary at War, being produced, were carried to Lord Cornbury, who at once gave directions for the regiments to march at five o'clock towards the enemy. Before daylight accordingly, on the 12th, the troops marched, and "continuing through that day and the following night, on the afternoon of the 13th they arrived at Axminster, within six miles of the Prince of Orange's head-quarters, where they were joined by the Earl of Abingdon, Sir Walter Clarges, and about thirty other gentlemen who pretended to be volunteers. It was now given out that a design of the Dutch to surprise the King's forces had been discovered, and orders were issued for beating up the quarters of the enemy that same night, and the three regiments were again in motion until they were met by a large body of cavalry which the Prince of Orange, apprised of their approach by Lord Cornbury, had sent forward. The greater part of the men, however, on becoming aware of what was taking place, and resolving not to join the Prince, galloped back. Major Robert Clifford of the Royal Dragoons brought off that regiment, with the exception of a few officers and about eighty dragoons, who accompanied Viscount Cornbury. The Blues also returned, excepting about twenty-seven, but the Duke of St. Albans' regiment, the 8th Horse, having mustered at a distance, the men, ignorant of the transaction, followed Lieutenant-Colonel Langston to Honiton, where they were welcomed as friends by the Dutch general.[1] Many of the men, however, returned to the royal service, and the Duke of Berwick having collected the remains of the three regiments, marched them back to Salisbury.

On the 20th of November King James arrived at Salisbury, where his Majesty rewarded the loyalty of Major Clifford by promoting him to the colonelcy of the Royal Dragoons on the 24th of the month, *vice* Lord Cornbury. The King, however, soon discovered that the defection among the officers was general, and that the soldiers, although reluctant to desert his service, were ill disposed to fight in the cause of Popery. The superior officers of the army with the nobility and gentry continued to flock to the Prince's standard, until James, alarmed for his personal safety, returned in haste to London, the Royal Dragoons at the same time moving into garrison at Portsmouth. The Prince of Orange advanced to the capital without serious opposition, arriving at St. James's on the 18th of November, when, James having fled to France, he assumed the government, and on the 30th of December his Royal Highness reappointed Viscount Cornbury to the colonelcy of the Royal Regiment of Dragoons, which went into quarters at Farnham and Alton.[2]

The crown was now conferred upon William and

[1] Lingard's *History of England.*
[2] *London Gazette*, *War Office Records*, *Life of King James II.*, &c.

Mary, Prince and Princess of Orange, who were crowned on the 10th of April, 1689.

Their Majesties' accession, however, did not pass without opposition, and Viscount Dundee having induced several of the Highland clans to take arms in favour of King James, the Royal Dragoons were immediately sent to the north,[1] and at the same time the Earl of Clarendon declining to act with the new Government, his son Lord Cornbury was superseded in the colonelcy of the regiment by Lieutenant-Colonel Anthony Hayford, whose commission as colonel is dated the 1st of July, 1689.

In this first year of the reign of William and Mary appeared the earliest promulgation of the Mutiny Act and the Articles of War.

On the 27th of July six battalions of infantry and two newly raised troops of Scots horse, commanded by Lieutenant-General Mackay, were defeated at Killiecrankie by the Highlanders and a few Irish, under "Claverhouse" Viscount Dundee and Brigadier-General Cannon, the former of whom was killed in the action; immediately after which the Royal Dragoons being ordered to march to the assistance of Mackay, they arrived at Perth in the early part of August. The object of the commander-in-chief being the prevention of the mountaineers from a descent into the Lowlands the regiment was posted for a short time at Forfar, under Major-General Sir John Lanier, and thence proceeded by forced

[1] *War Office Route Book.*

marches to Aberdeen. The Highlanders eventually retired over the mountains by paths inaccessible to cavalry, and separated to their homes.

Meanwhile the Lord-Lieutenant of Ireland, the Earl of Tyrconnel, having retained the greater part of that kingdom in the interest of James, who landed at Kinsale on the 13th of March, 1689, from Brest, King William sent thither Marshal Schomberg, who landed at Carrickfergus with 16,000 men on the 13th of August, and in the beginning of October the Royal Dragoons being ordered upon this service, on the 9th of that month[1] they landed at Carlingford and proceeding to Armagh and Clownish they moved thence to the Isle of Maghee.

Some skirmishing occurred during the winter, and in the spring of 1690 the regiment was before Charlemont, which place was blockaded by the King's forces and defended by a garrison of 500 men commanded by Sir Teague O'Regan, who surrendered on the 14th of May, when a detachment of the Royal Dragoons escorted the garrison to Armagh.

In June following Lieutenant-Colonel Edward Matthew, from "Leveson's" Dragoons, now the 3rd King's Own Hussars, was appointed colonel of the regiment which was encamped at Loughtrill, where it was joined by a remount from England.

On the 22nd of the month King William arrived at the camp, and "his Majesty was no sooner come than he was in among the throng of the troops and observed

[1] *London Gazette.*

every regiment very critically. This pleased the soldiers mightily, and every one was ready to give what demonstrations it was possible, both of his courage and duty."

His Majesty had landed at Belfast on the 14th of June, and on the 30th he appeared with 36,000 men on the banks of the Boyne, of which river James with an army of Irish and French prepared to dispute the passage.

The celebrated battle of the Boyne was fought on the 1st of July, which ended in the complete rout of his army and the ruin of the cause of James, who fled by way of Dublin to Waterford, and thence to Brest. The Royal Regiment of Dragoons, commanded by Lieutenant-Colonel Edward Leigh, with the other troops engaged at the passage of the Boyne, are reported to have "acquitted themselves well;" and the army of William advancing to Dublin, his Majesty reviewed the regiment at Finglass, where it brought 406 sabres into the field.

On the 21st of July Major-General Kirke proceeded with the Royal Dragoons, the Queen's Dragoons and Colonel Cambron's regiment of foot to Waterford, where he summoned the place, which capitulated on the 28th.

At this moment, while success was thus attending the operations of the army in Ireland, the English and Dutch fleets, commanded by Admirals Byng and Evertsen, were defeated in an engagement off Beachy Head by the French, under the Count de Tourville, and, the enemy threatening a descent on the British coast, William ordered a troop of the Life Guards, Count Schomberg's "Horse," now the 7th P. R. Dragoon Guards, the Royal

Dragoons, with Trelawny's and Hastings's regiments of foot, now the 4th and 13th of the line, to embark immediately for England.

The Royal Dragoons landed at Highlake, in Cheshire, in the early part of August, 1690, but the alarm of invasion soon subsiding, they returned to Ireland, where, landing on the 20th of October, they took up extensive cantonments in the county of Cork. Many thousands of the Roman Catholic peasantry were at this period in arms for King James, who, forming themselves into bands called "rapparees," made frequent incursions into the quarters of the English regiments. Several men of the Royal Dragoons were murdered in their quarters, and detachments were constantly employed in scouring the country and chasing the "rapparees."

Towards the end of December a detachment of the regiment proceeded with some other troops on an expedition commanded by Major-General Tattea, which on the 1st of January, 1691, attacked a fort near Scronclaird, and took it in two hours, although the Irish had employed 800 men during two months in the construction of it.[1]

In the spring, when the army took the field, the Royal Dragoons remaining in the county of Cork, in the early part of June, Major Calliford with a detachment of the regiment and some militia penetrated into that part of the country whence the enemy drew their supplies, defeated their troops and captured several droves of cattle, until at length General Ruth, who commanded the

[1] Story.

French and Irish forces, detached 2,000 horse and foot to cover that neighbourhood. Major Calliford, however, persevered in making inroads, and having advanced upon one occasion with 120 of the Royal Dragoons and 56 militia he fell in with two troops of Irish cavalry. The English dragoons made a bold charge upon their opponents, killing twenty on the spot, and pursued the remainder to Newmarket, where the Irish, being reinforced, made another stand. The Royals, however, attacked them again with great bravery, and having killed eighteen, the rest fled in disorder, leaving behind a quantity of provisions and some cattle. Major Calliford despatched 11 dragoons and 24 of the militia to the rear with the booty, and then pursued the fugitives four miles further, when he came upon a body of 500 of the enemy's horse, commanded by Sir James Cotter. Notwithstanding the great disparity of numbers the Royals showed an undaunted front, but were at length overpowered with the loss of 40 men, when Calliford made good his retreat with the remainder. On retiring, the dragoons, chafing and burning for revenge, frequently turned round upon their pursuers, until at length Captain Bower and twenty men faced about and killed nearly twenty of the Irish, whose eagerness in the chase had carried them in front of their main body. Meanwhile, the party detached with the captured stores and cattle arrived at Drumaugh, where, being attacked, they defended themselves with success until relieved by some troops under Colonels Hastings and Ogleby.

While the Royal Dragoons were thus engaged the army commanded by Lieut.-General de Ginkell gained a decisive victory over the French and Irish on the 12th of July at Augrim, in which affair General St. Ruth was killed, and on the 18th of August the regiment joined the army at Banagher Bridge. The enemy having collected the remains of their defeated regiments at Limerick, De Ginkell undertook the siege of that city on the 26th of August, commencing operations on the right bank of the Shannon; the Irish army at the same time lying encamped on the opposite side of the river.

A pontoon bridge having been prepared, at daybreak on the 16th of September several regiments were ordered to cross the river, the Royal Dragoons taking the lead, when Brigadier-General Clifford,[1] formerly Colonel of the regiment, but now commanding four regiments of King James's dragoons, being taken by surprise, made little opposition; some infantry, however, attempted to make a stand, but a squadron of the Royals dashing forward routed them in an instant. Two or three French and Irish battalions took refuge in a bog and wood in their rear whence they were driven with the loss of several killed, and of a French lieut.-colonel, a captain, and a number of men taken prisoners. The regiment, which had passed the Shannon, fell upon the enemy's camp, where they found a scene of the greatest confusion, many

[1] Colonel Clifford, of the Royal Dragoons, adhered to King James at the Revolution, and having proceeded to Ireland he was appointed a Brigadier-General.

of the Irish running about in their shirts; pulling down tents; many making their escape into the city, while others fled towards the mountains. A regiment of dragoons, whose horses were at grass two miles away, dispersed in confusion; while a party of Horse, taking to their arms, made a show of resistance, but made off on the approach of the English, who took possession of the camp, in which they found a quantity of beef, brandy, and corn, together with the saddles and appointments of 300 dragoons. The Royal Dragoons were commended by Lieut.-General de Ginkell for their conduct, and the same day they returned to their own side of the river.[1]

On the 22nd of the month the regiment, with several other corps, crossed the Shannon into the county of Clare, when the advanced guard, consisting of eighteen men of the Royal Dragoons, was attacked by a squadron of Irish cavalry, whose first onset they met with admirable resolution, but were forced to retire, until part of the regiment coming to their assistance the enemy were defeated and chased under the range of their batteries with the loss of three small pieces of brass ordnance.

Orders were now given for the infantry to attack the works covering Thomond Bridge, which being carried after a severe struggle, their defenders endeavoured to escape into the city, but the drawbridge having been raised, they were left to the mercy of the besiegers, who

[1] Story.—*London Gazettes*, &c., &c.

slaughtered them in such numbers that the slain lay in heaps on the bridge higher than the parapet. Five colours were taken, and so many were killed, drowned, and made prisoners, that on the 3rd of October the place surrendered and the rebellion in Ireland was at an end.

CHAPTER II.

RETURN TO ENGLAND.

In January, 1692, the Royal Dragoons returned to England, and went into dispersed cantonments in Leicestershire, a detachment during part of the summer being in garrison at Portsmouth. The regiment was subsequently employed on revenue duty in the maritime towns on the south coast, and in the autumn of 1693 it had the honour of furnishing escorts to attend King William from Margate to London on the return of his Majesty from Holland.

The war with France, commenced in 1689, had been continued with varied success, and in the spring of 1694, the Royal Regiment of Dragoons being ordered on foreign service, they embarked in May, joined the army encamped near Tirlemont in South Brabant on the 21st of June, and were reviewed by King William on the following day. On arriving at the camp they were posted in front of the village of Camtich, which quarter being much exposed, they were reinforced by two regiments of Dutch infantry. The army marched from Tirlemont on the 13th of July and encamped at Mont St. André and Ramillies, where the regiment was brigaded with the Royal Scots and Fairfax's Dragoons, now the Scots Greys, and the 3rd King's Own Hussars, under Brigadier-General Mathews, whose brigade was encamped on the left of the line. The French lay encamped near Huy, with their left upon the Mehaine. On the 17th of July a foraging party of the allies crossed that river, and meeting with several French squadrons, a skirmish ensued in which the Royal Dragoons lost eight horses and had three men wounded. On the 23rd another party encountered a detachment of the enemy, when the regiment had two men and several horses killed. The allied army was again in motion on the 3rd of August, when much manœuvring and some skirmishing took place, but no general engagement. On the 29th the Royal Dragoons were stationed at Wacken, situated at the junction of the Mandel and the Scheldt, whence in October they moved into cantonments in villages between Ghent and Sans-van-Ghent.[1]

In the spring of 1695 the Royal Dragoons marched to Dixmude, forming part of a division of the army commanded by Major-General Ellenberg, and were brigaded with Lloyd's Dragoons, now the 3rd King's Own Hussars, and a regiment of Danish cavalry. On the 7th of June the Duke of Wirtemberg took command of this division, and attacked the French forts at Kenoque as a diversion to conceal King William's design upon the important and almost impregnable fortress of Namur, which

[1] D'Auvergne's *History of the Campaigns in Flanders.*

was invested shortly afterwards. The Royal Dragoons joined the covering army towards the end of June, but in July they were detached to Bruges, whence they were recalled to the camp between Genappe and Waterloo, proceeding thence to the vicinity of Namur. After the surrender of that place by Marshal Boufflers, on the 5th of September, they went into cantonments behind Ghent.

In the spring of 1696, the French threatening an attack upon the allied quarters in Flanders, the regiment was suddenly called from their cantonments to encamp upon the banks of the canal between Ghent and Bruges, where, on the 29th of May, they were reviewed by King William. They served the campaign of this year with the army of Flanders, commanded by the Prince of Vandemont, and were brigaded with the Royal Scots, the Royal Irish Dragoons, the present Scots Greys, and the 5th Royal Irish Lancers, under Brigadier-General Mathews. The object of this army was the protection of Ghent, Bruges, and the maritime towns of Flanders. No general action occurred, but a party of the Royal Dragoons, with one of Langston's Horse, now the 4th Royal Irish Dragoon Guards, surprised one of the French pickets on the night of the 20th of September, and took thirty prisoners. This appears to have been the only occasion on which the regiment was engaged during the campaign of this year; and on the 6th of October it went into quarters in the villages behind the canal of Bruges.

Throughout the campaign of 1697 the regiment served

under King William in the army of Brabant, and was brigaded with the Royal Scots and Eppinger's Dragoons, a foreign corps in British pay. On the 28th of May Brigadier-General Mathews died, and on the 30th his Majesty conferred the colonelcy of the Royal Regiment of Dragoons upon Thomas, Lord Raby, afterwards Earl of Strafford.

The enemy, having great superiority of numbers, besieged and captured Aeth, and afterwards threatened Brussels, but were frustrated in their designs by King William. The Royal Dragoons encamped before Brussels in June, and subsequently at Wavre; and hostilities terminating on the 30th of September by the treaty of Ryswick, the regiment embarked from the Netherlands, and landing at the Red House in Southwark on the 21st of November, at the end of the month it moved into Yorkshire, when the establishment, which during the war had been eight troops, amounting to 590 officers and men, was reduced to six troops of 294 officers and men.

During the two succeeding years the regiment occupied quarters in Lancashire and Leicestershire. In June, 1700, it was reviewed on Hounslow Heath by King William III., who was pleased to express his approbation of their appearance and discipline; and in the month following it moved into Yorkshire and Cumberland, with one troop at Carlisle and another at Hull.

King Charles II. of Spain dying on the 1st of November, 1700, Louis XIV. of France, regardless of former treaties, put forward the claims of his grandson Philip, Duke of Anjou, to the vacant throne, and in view to hostilities, the Royal Dragoons, augmented to eight troops of 532 officers and men, embarked for Holland in the beginning of March, 1702. Before the transport sailed, however, William III. died, on the 8th of the month, and the regiment was disembarked and placed in cantonments in the immediate vicinity of London. In a few days afterwards, Her Majesty Queen Anne resolving to pursue the foreign policy of her great predecessor, the regiment was re-embarked, and landing at Williamstadt, went into quarters at Breda, where it was again brigaded with the Royal Scots and the Royal Irish Dragoons, under that excellent officer, Brigadier-General Ross, and was employed as a guard to the English train of artillery.[1]

The formal declaration of that war, which is known in history as the "War of Succession in Spain," and which more or less embroiled the whole of Europe for nine years, was made on the 18th of May, 1702, in London, Vienna, and the Hague. The league against France and the Duke of Anjou, and in favour of the Archduke Charles of Austria, comprised England, Holland, Savoy, Austria, Prussia, and Portugal, while Spain and Bavaria supported the cause of Philip.

The war was to be carried on upon four separate theatres—Belgium, the valleys of the Middle Rhine and the Upper Danube, the Sierras and coast of Spain, and the North of Italy.

[1] *Official Records*, *London Gazettes*, &c., &c.

A powerful French army was in the field threatening the frontiers of Holland. The Duke of Marlborough assembled his forces towards the end of June, and in July the Royal Dragoons joined them with the artillery. They were then employed in covering the sieges of Venloo, Ruremonde, and Stevenswaert, and took part in the capture of the city of Liége, afterwards returning to Holland to be quartered at Arnheim, the capital of the province of Guelderland, where, in April 1703, they were reviewed by their colonel, Lord Raby, who was passing through Holland on his way to Prussia as envoy extraordinary to that court.[1]

At the commencement of the campaign of 1703 the regiment was engaged in covering the siege of Bonn. Thence, on joining the army near Maestricht, with six battalions of infantry, commanded by the Prince of Hesse, it was brigaded with the same corps as in the preceding year.

On the advance of the Duke of Marlborough's army the French retired, and took post behind their fortified lines between Camphont and Westdown, towards which, on the 27th of July, the British commander proceeded with 4,000 horse and dragoons, when Lieutenant Benson, with the advanced guard of 30 men of the Royal Dragoons, charged and overthrew a picket of 40 French horse, and chased them to the barriers of their entrenchments, thus giving his Grace an opportunity of approaching within musket shot of the lines

[1] *London Gazettes*, *Millner's Journal*, and *Annals of Queen Anne.*

which he was desirous of attacking, but was prevented by the timidity and pertinacity of the Dutch generals and field deputies.

At the siege of Huy, which was invested on the 16th of August, the Royal Dragoons were encamped on the river Maese, in order to secure the bridge and keep up the communication; and were subsequently employed at the siege of Limburg, which was invested on the 10th of September, a city upon a pleasant eminence among the woods near the banks of the little river Wesdet, which surrendered on the 27th of the month. Spanish Guelderland being now delivered from the power of France, and the Dutch freed from the dread of invasion, the Royal Dragoons returned to Holland, while in the meantime circumstances had occurred which occasioned their removal from the army of the Duke of Marlborough to another theatre of action.

It was now determined by the British Government to send a force to Portugal in support of the claims of the Archduke Charles in the Spanish Peninsula; and the Royal Dragoons being included in this force, they gave up their horses to the number of 397 to the British regiments remaining in Holland, on the 6th of October, 1703, and on the 10th they embarked, dismounted, at Williamstadt in the vessels that were to take them to Lisbon, the strength of the regiment being 403 of all ranks.

At this period there was serving in the Royal Regiment of Dragoons as captain, and afterwards as major and brevet lieutenant-colonel, a French officer, Lieutenant-Colonel de St. Pierre, whose *Journal*, commencing in the month of October, 1703, and continued to the end of 1706, has been discovered, and published by Major-General E. Renouard James, late of the Royal Engineers, and contains many details little known both as regards the regiment and the "War of Succession in Spain," of which the *Memoirs* of Captain Carleton, published in 1728, and re-issued in 1808, and the *History* of Lord Mahon, in 1832, are the only standard accounts. It may be interesting to learn that Lieutenant-Colonel de St. Pierre was a collateral descendant of that Eustache de St. Pierre so memorable as the defender of Calais when besieged and taken by King Edward III. in 1347. He had also in the regiment a brother-in-law, Lieutenant Peter Renouard, who was aide-de-camp to General Windham in Holland, and subsequently in Spain to the Earl of Peterborough, to the Archduke Charles of Austria, one of the rival claimants, and to Lord Galway; and who, in 1707, purchasing the Majority of Brown's regiment of Horse, of which he became Lieutenant-Colonel in 1732, retired from the service in 1743, having served twelve campaigns abroad and one in 1718 in Scotland. Lieutenant-Colonel Renouard died in Dublin in 1762, at about eighty-two years of age. Both he and De St. Pierre were descendants of Huguenot refugees after the Revocation of the Edict of Nantes in 1625; and it may be added that they were both equally ancestors collaterally of the General James by whose publication of the *Journal* in question the records of the Royal Regiment of Dragoons will now so greatly benefit.

Lieutenant-Colonel de St. Pierre thus prefaces his *Journal* :—

"Règles que doit obseruer un historien :—beaucoup d'ordre ; un style net, court, simple, sans affectation, sans figures, ny autres ornements oratoires ; une grande sobriété dans l'éloge et dans le blâme des différentes parties, soit de politique, soit de religion."

Having therefore embarked at Williamstadt on the 10th of October, St. Pierre goes on to say that :—

"Being unable to put to sea in consequence of contrary winds, there came on in the meantime one of the most terrible storms ever known, in which sixteen men-of-war were lost with all on board, and, in one of these, Rear-Admiral Beaumont. The three transports in which were embarked the Royal Dragoons ran all ashore, and were in great danger of being lost."

And further on :—

"Little care was taken of the forces that were about Helvoet Sluice, no shipping nor shift being made for those that wanted it. Sir George Rooke sayled away for England with the King of Spain and a fair wind, from ye Brill, on Friday the 4th January, 1704.

"Stewart's regiment was the only one that went

entire, and the Royal Dragoons the only one that remained here entire.

"After the storm many men of the regiment fell sick. Most of them were putt in a church or a vestry att a place called Old Bone, and for all the care that was taken of them several dyed. It is remarkable that the regiment had but one man sick when we embarked."

On the 5th of September, 1703, died at Breda Colonel Rossiter, a circumstance thus noted in the *Journal* :—

"Oct. ye 6th.—Richard Rossiter, Major of the Royal Regiment of Dragoons, dyed at Breda after a long sickness; he came into the regiment cornet in ye year 1685; was made a captain in 1689, major in the year 1697, had a brevet as lieut.-col. in ye year 1703, in May.

"About that time we had news from England that St. Pierre was made major of the regiment in the room of Rossiter. Captain Killigrew had the said Rossiter's troop, and Captain John Wyvill bought Captain Young's troop for £7,000. Shelden sold his place of captain-lieut. to Sam. Jason; Lieut. Richardson sold his place of lieut., and John Farnham succeeded him in Croft's troop. Best bought cornet in his room; Charles Harris was made lieut. to Col. Killigrew; John Pulford was made cornet in Captain Killigrew's troop; Cornelius Tylbury was made cornet in Wyvill's troop; and Tattershall sold his quarter-master's place to John Topham."

The continued severity of the weather and the disasters caused by the storm prevented the ships from putting to sea, and it was "judged impossible by the pilotes of the men-of-war and of the Brill to put to sea without great danger of losing the ships by reason of the ice." The troops embarked, suffered greatly from the cold; and St. Pierre tells us that on—

"Tuesday ye 18 of January, 1704, great application had been made by Major St. Pierre to my Lord Cutts in case they were frozen in, to gett quarters for the men; and he had sent Cornett Renouard on purpose to the Hague with the application made to my Lord Cutts, and by him to the States; but one Sadler that was employed about the transport, telling the said Lord Cutts that wee were going to sail, he left off the sollicitation upon the supposition wee were gone. We remained in a distressed condition: no money, not above three days' provisions, the rivers frozen, and a great many men sick. Some boats from Rotterdam came and brought us provisions. The States sent three Commissioners of their body to the Brill, who sent for all the commanding officers at Helvoet Sluice to come there at the Brill, which they did the 26 of Jan. They offered us, in the name of the States, shipping, provisions, quarters, money, and whatever we wanted. We took a little money, £620, for ye regiment; and the frost being gone and the weather opened, we refused quarters. My Lord Duke of Marlborough landed the next day at Helvoet Sluice. He saw some of the men there, and found them shrunk much. Some time after application was made to him by M^r^. St. Pierre to desire the men should be putt to full allowance, upon which my lord writt to Captain Atkinson to do it, if it was possible without any prejudice to ye service, being very much as he thought for her Majesty's service, and a means to recover the sick and strengthen the rest. Captain Atkinson agreeing with the reasonableness of the thing, answered my lord by letter, and promised that wee should be putt at full allowance if wee sayled soon with a fair wind. He gave good words to my Lord Duke, and to us, and did nothing."

"The *John and William* being gott off from the bankside and refitted, it was given again to the regiment; St. Pierre and Wyvill put two troops on board of her, and wee sayled for England with a fair wind Saturday ye 23 of Feb."

"Came to an anchor at Spithead, Tuesday ye 26 Feb."

"The King of Spain sayled from Spithead Sunday ye 23d, and with him Captain Killigrew, Lieut. Farnham, Cor. Best, Cornet Tyboury, G^l^.-M^r^. Lightfoot and 113 men, recruits of the regiment."

"*Tuesday, March ye 4th.*—Sett sail from St. Helen's with a very fair wind under the convoy of ten men of warr Dutch and English commanded by Vice-Admiral Sir John Lake, six of ye men of war parted with us off the Burlinges, and went some to ye east and some to ye west Indies."

"*Monday, ye* 12 *of March* 1704.—Came to an anchor before the castle of Lisbon, where the King of Spain with the squadron under Sir George Rooke's command was come five or six days before with the army on board, of which none were yet landed, and began only to land about ye 16 or 18."

The allied force sent to Lisbon was composed of 4,000

Dutch under General Fagel and 6,000 English under the Duke of Schomberg. The Archduke Charles had been received with great honour on his arrival there on the 6th of March, but no preparations had been made for the approaching campaign. The army of Portugal was thoroughly inefficient; the fortresses were in a dilapidated condition; great difficulty was experienced in mounting the English cavalry, which had been guaranteed by the Portuguese Government, and all difficulties were increased by Fagel being on bad terms with Schomberg.

On the other hand, the state of affairs in Spain was very different, where the Duke of Berwick, a nephew of the Duke of Marlborough, and a man of great military talent, was in chief command. In addition to the Spanish troops, his force was increased by 12,000 auxiliary French, the numbers in all amounting to 35,000 men with reserves. Philip of Anjou, the rival of Charles, accompanied the army in person, and the most careful preparations had been made.

The circumstances attending the arrival of the Royal Dragoons at Lisbon are thus related by Lieut.-Col. de St. Pierre:—

"Brigadier Hervey, as Brigadier, pretended to command both regiments, and promised to take as much care as possible could be of ours. He landed his about ye 18th, and put them in the barracks at Alcantara, and left us to shift for ourselves; if these barracks had been well divided there had been near room enough for both regiments. The officers of the said Royal Regiments complained highly of such partial proceeding; but all in vain; they offered them an empty monastery, called Mestera de Rates, which had been a good place indeed if the best and greatest part of it had not been appointed for a general hospital for the English. Indeed at that time there had been room enough for both, but the officers did not care to putt their well men among the sick men of the army.

"Other disputes arrose with the same B. Hervey about choosing of the horses. He would not allow the said regiment to choose horses equally with his own, because they were Dragoons, though they were told by the General before the said B. they were to charge as horse, and to pay the same price for the horses as they did. The B. interest carried it against the good right of the said regiment; and he had an order from the Duke to choose the horses of his regiment before the Royals; he disputed all amounts with the said regiment, and went so farre, the question being asked him, if the two regiments were to quarter in the same place, and there was room but for one, if he would take all the stables and quarters and order us to encamp. He said he would; and that he never would allow us to draw or roll with his regiment, notwithstanding wee produced an order given by King Charles ye Second, which giveth post to ye Dragoons of all those raised after them, specially to our regiment."

"*March ye 24th.*—No quarters being to be had, notwithstanding all our sollicitations, the Regiment was ordered to land, and encamped about Belle Isle, four miles from Lisbon, cold and bad refreshment enough for people that had been near five months on shipboard, specially the weather proving rainy, stormy and cold enough for the country they were in.

"No troops, I believe, were ever more neglected, notwithstanding the daily clamours of the officers; who would not have thought that there had been good quarters provided for troops that had been so long at sea and suffered so many storms: who would have thought that our generals would have suffered them to be landed and to encamp them and to pin up the basket for near a fortnight. It proved to be extraordinary rainy weather; men fell sick every day. Bad weather, scarceness of victuals, and plenty of wine were the chief causes of it. It was at first landing, towards the end of Lent, for that reason, or for not being used to it, the Portuguese brought little or nothing into the camp, and what they brought was extraordinary dear and butt very indifferent."

On the 8th of April the Royal Dragoons were ordered into the castle of Lisbon, and the mounting of the regiment promised by the Portuguese Government, but very ill performed, was now commenced by Major de St. Pierre, who writes:—

"In the meantime I was employed to chuse horses with Baron Winterfield, Colonel of a Walloon Regiment of Dragoons. In four or five days time wee choose 169, which wee divided between the two regiments very amiably, and amongst them together with fifteen that had been chosen for the officers. They were divided in the manner following:—

My Lord's	Troop . . .	1 Officer's horse, 10 men's.
Colonel Killigrew's	,, . . .	2 Officers' horses, 10 men's.

Colonel St. Pierre's Troop	. . 3 Officers' horses, 11 men's.
Captain Graves's ,,	. . 3 Officers' horses, 10 men's.
Colonel Croft's ,,	. . 2 Officers' horses, 10 men's.
Captain Peake's ,,	. . 2 Officers' horses, 11 men's—1 being for his Lieutenant and 1 for his Quarter-Master.
Captain Wyvill's ,,	. . 3 Officers' horses, 10 men's—1 being for his Lieutenant.

"*April ye 12th.*—I received from England a Brevett of Lieut.-Colonel, dated 1st day of the year 1704, and that same day I drew out for the first time about 80 men that had been mounted of the horses I had chosen, which came to exercise much better than I expected."

This mounted detachment proceeded under Captain Peake to the frontiers of Portugal, and encamped on a pleasant plain near Estremos, and while there it accompanied an expedition into the Spanish territory, which, under Dom Joan de Lancaster penetrated as far as Olivenea and Barcarola, where they ordered the proclamation of Charles as King of Spain in the church and market-place. The remainder of the regiment meanwhile proceeded, in June, dismounted, to Abrantes, there to await the arrival of horses, and on the way, at Piquette, Lieut. Farnham going to bathe, was unfortunately drowned.

On the 7th of July the regiment went into quarters at Abrantes; on the 23rd of which month Gibraltar was taken by the combined fleets of Sir George Rooke and Sir Cloudesley Shovel, and 2,000 marines commanded by the Prince of Hesse Darmstadt.

On the 7th of August St. Pierre tells us that,

"116 extraordinary bad horses were given by the King, for which wee were told the Queen not wee were to pay. Fourteen were given to each troop, and six to ye hotbois and kettledrum. Captain Peake fell sick as soon as he came into the town with the party from the *Alantejo* of a fever and bloody flux. About that time wee heard that the Duke of Schomberg was recalled, and that my Lord Galloway was coming to command in his place, which caused a universal joy to the whole of the army by the just opinion they have of their new general, who landed with a small attendance at Lisbon on Tuesday, August ye 10.

"Captain Peake's sickness continued, and att last he dyed, August ye 14th. He was a young gentleman of twenty-two years, endowed with a great many good qualities, handsome in body and of very clear understanding, which had been much improved by his being bred in the university. He applyed and delighted much in souldering, and if he had lived he would have made as good an officer as any in the kingdom. He dyed lamented by every one that knew him, and was buryed on Friday ye 18th in the castle, with small ceremony. Eighty dragoons, with their arms in the funeral posture, were led by two quarter-masters. The men marched four and four, and just before the last rank, that was composed of corporals, marched four drums, their cases covered with black, next to them the corporals, and afterwards two cornetts carrying the standards, in a funeral posture, with black to each standard; then two lieuts., then the captain that commanded the party, after him the hotbois playing a doleful tune, which were followed by the surgeon and the chaplain and the corpse, which was attended by the colonel and a great many other officers. As usual, three volleys were given."

"*Tuesday, Feb. ye 22nd.*—Colonel Killigrew marched with the mounted men of the Regiment upon a pressing order from My Lord Galloway to join the army with all expedition, and left me here with the men on foot, the number being as follows—each troop one sergeant, two corporals, one drum :—

My Lord's	Troop	20 men.
Lieutenant-Coll.'s	"	20 "
Major's	"	17 "
Captain Greaves's	"	20 "
Captain Croft's	"	18 "
Captain Pelle's	"	19 "
Captain Killigrew's	"	19 "
Captain Wyvill's	"	20 "
		153 "

Remained in Abrantes, 1 sergeant, 1 corporal, 1 drum of each troop, and private men well and sick.

My Lord's	Troop	28 men.
Lieutenant-Coll.'s	"	21 "
Major's	"	24 "
Captain Graves's	"	18 "
Captain Croft's	"	16 "
Captain Pelle's	"	19 "
Captain Killigrew's	"	21 "
Captain Wyvill's	"	18 "
		161 "

"At a court-martial held at Abrantes, Saturday, ye 21st of 6re, Coll. Allen, President, Cambell of Captain Graves his troop, was accused of having offered to draw his sword against Sergeant Carr. The court left him to be punished according to ye discretion of the commanding officer.

"Advice being taken of the officers, I had him whipt and turned out of the Regiment. He went into Stewart's regiment and soon after was hanged for robbery and murther."

On Friday, the 22nd of November, the dismounted men, under Lieutenant-Colonel de St. Pierre, quitted Abrantes, and the whole regiment went into winter quarters in the villages in the Alantejo.[1]

The state of the Royal Dragoons at this time, from Colonel de St. Pierre's notes, was as follows:—

"*Thursday, ye* 11*th of Xbe.* 1704.—I went to visit the several troops of the regiment, and had the articles of warr read to them. I found—

My Lord's Troop, sick and well . .	Private men	41	no drums.
Colonel Killigrew's I did not see.			
My own Troops, sick and well . . .	"	38	
Captain Graves's " . . .	"	36	
Colonel Croft's " . . .	"	32	
Captain Bensen's " . . .	"	35	
Captain Killigrew's " . . .	"	37	
Captain Wyvill's " . . .	"	22	
Sergeants and Corporals compleat.			

Assuming the troop not seen as of the average strength, and making due allowance for Non-Commissioned Officers not here included, the Regiment would seem to have numbered about 300 men. We know there were very few horses with it."—E.R.T. EDITOR'S *Note.*

In April, 1705, the Royal Dragoons advanced with

[1] *London Gazettes, Present State of Europe, Mémoires de Berwick, Annals of Queen Anne*, and *Official Records* in the War Office.

the army into Spanish Estramadura, and were present at the capture by storm of Valencia de Alcantara, on the 9th of the month, and of Albuquerque, whence they proceeded on the 18th of May to St. Ubes, and thence to Lisbon, there to receive horses sent out from Ireland, and a batch of recruits; the former being distributed thus—120 to Hervey's Regiment of Horse, now the 2nd Dragoon Guards, or Queen's Bays; 400 to the Royal Regiment of Dragoons; 300 to Cunningham's Dragoons, now the 8th R. I. Hussars; and 431 to Winterfield's Dragoons.

Meanwhile in England another expedition had been fitted out in aid of the cause of Austria, of which the land force was commanded by Lieutenant-General the Earl of Peterborough, who, arriving at Lisbon at the end of June, the expedition into Catalonia and Valencia was promptly decided on, in which the Royal Dragoons, with Cunningham's, and four regiments of Foot, were included, and this force the Archduke Charles resolved to accompany in person.

On the 28th of July, 1705, the regiment embarked at Alcantara, the Archduke, five days previous, having gone on board the *Ranelagh*, his determination to accompany the troops being on several accounts by no means agreeable to Lord Peterborough, but which could not with propriety be refused.

St. Pierre writes :—

"Wee embarked 37 men from every troop, and all

the officers but Captain Killigrew, Lieut. Topham, and G. M. Donington.

"The fleet set sail from Lisbon July ye 28, 1705, with a fair wind. God grant us good success.

"By the fault of the master, who had in the evening the fleet in sight at the head of him, and who lay by all night, thinking the fleet would do the same, wee left the said fleet, and were obliged to come without convoy as far as Gibraltar, in company with another transport, and we arrived there safely.

"The fleet being all joined there, and having landed Elliott's and Enfield's Regiments that were to remain there, wee took in their room all the marines; the battalion of Guards, the Regiments of Barrimore, Moncy, and Donegall; and the Prince of Hesse went on board of the *Namur*.

"We sailed from Gibraltar up the straits, along the coast, and came to Altea Bay in the kingdom of Valencia, where wee halted for four or five days near a little town called Altis (Altea), where the country people flocked in great numbers, and where upwards of ten thousand came on board the *Britannia* to kiss the King's hand, who promised to sett up for him. Some of these performed their word soon after."

Leaving Altea Bay, on Sunday, the 9th of August, the fleet continued on towards Barcelona, coming in sight of the city on Saturday, and passing on, they came to anchor above it, and on the east side, when the infantry began to land on the 22nd. The Royal Dragoons disembarked on the 24th, and encamped near a river called Bassoz, on the east side of the city and about a mile from the walls.

The siege of Barcelona was considered a romantic enterprise, which excited a lively interest in every nation in Christendom. The garrison equalling in strength the besieging army within about 2,000 men, success, according to the ordinary rules and chances of war, appeared impossible. The siege, however, was commenced, and on the 14th of September the strong fortress of Montjuick, built on the loftiest of a cluster of heights in the immediate vicinity of the city, which it screened from approach on the inland or western side, was attacked and taken after a resistance of three days, which success, however, cost the life of the Prince of Hesse, of whom St. Pierre thus speaks:—

"The Prince of Hesse dyed within a few hours of his wounds, mightily lamented by all that knew him, for he had all the good qualities that a man could have to gain the affections of the people—handsome of his person, valorous, generous, and ready always to do good, but especially the Catalans had a singular estime and veneration for him. He was a younger brother of the house of Hesse Darmstadt. He served first amongst the English; and had an English regiment of Foot given him. He was at the battle of the Boyne with King William, but afterwards the King of Spain having married a sister of the Elector Palatine, a near relation of the Prince, he went into Spain, changed his religion; was made General of the Horse, and behaved himself very well during the siege of Barcelona in the year

1692. He made several salyes, in which he did the French a great deal of mischief, and got mighty reputation. After the peace was concluded he was made Viceroy of Catalonia; and there it was that by his sweet temper and just and moderate command he won the hearts of people in such a manner that after his death every one mourned as if it had been for a father.

"King Charles being dead, and the Duke of Anjou having taken possession of the crown of Spain, he retired into Germany, from whence he came along with King Charles, who gave him the title of Vicar-General of Arragon. It was an extraordinary loss the King had in him. In that juncture of times being certainly the greatest and most useful man that was come out of Germany with him."

The fall of Montjuick led to that of Barcelona, where the governor capitulated on the 9th of October. On the garrison preparing to march out, agreeably to the terms of the surrender, a serious insurrection broke out among the inhabitants, who attacked the houses of the French and others known to be in the interest of the Duke of Anjou, threatening even to massacre the Governor Velasco and the garrison. The Earl of Peterborough, however, marching in at the head of a troop of the Royal Dragoons and a detachment of Grenadiers, restored order and tranquillity, in doing which his lordship very nearly fell a victim to his humanity, for, while escorting the Duchess of Popoli, whose husband, a Neapolitan nobleman, was a lieutenant-general in the army, a ball fired by one of the rioters passed through the Earl's periwig.

By the conquest of Barcelona, at which, as Dr. Freind observes, "all Europe wondered," nearly every town in Catalonia declared for King Charles III., and St. Pierre says:—

"The troops expected to have some good quarters and some refreshment after their great fatigue. Instead of that, they were crowded into several baraques or convents, where they had no beds nor firing, nor any accommodation, notwithstanding the fair promises the gentlemen of ye country had made them to encourage them to go on with the siege. The officers having no quarters allowed, were fain to hire lodgings of the inhabitants, which would not let them under a year's time, and half of it in hand. Complaints were made of that ill-usage to the King; but as it was one of the privileges of the inhabitants not to quarter any souldiers, the King dared not oblige them to it, and their gratitude and generosity were not great enough to move them to it. In the meantime, the weather being something cold and rainy, no firing allowed them, lying upon the bare stones in the galleries of the convents, the men fell sick, and in a little time wee lost near the third part of our army. Att last it was resolved to send them into the country, which might as well have been done att first."

The Royal Regiment of Dragoons moved to Tortosa about the beginning of December, a very ancient town lying upon the river Ebro, near the frontiers of the kingdom of Valencia, which had declared for King Charles. The head-quarters of the regiment there, according to St. Pierre, consisted "of about two hundred men, very ill-mounted, the best mounted men having been detached under Captain Jasen to Lerida;" and there had been also a detachment left at Barcelona as a bodyguard to the Archduke Charles, who had assumed the title of King Charles III.

The relief of the town of St. Matthew, and the subsequent pursuit of the army commanded by the Conde de las Torres, forms one of the most remarkable episodes of the war, and of Lord Peterborough's marvellous energy, intelligence, and activity, and was commenced from Tortosa on the 1st of January, 1706, when he marched with three regiments of Foot, making about 1,100 men: 170 of the Royal Dragoons mounted upon horses "that could not have galloped a mile had it been to conquer the kingdom of Spain"; 150 Spanish dragoons newly raised, and without musquets; and was joined upon the road by 500 people of Vimaroz, with four pieces of cannon. Upon the approach of the Earl the enemy retired, his rear guard being pursued by the Royal Dragoons over the mountains to Albocazar, whence, continuing their retreat, it was followed up by Lord Peterborough with a force so inferior in numbers that the record of these events appears almost incredible,[1]

[1] "Notwithstanding King Charles has received no reinforcements since he landed in Catalonia, his partisans, and the small army under the Earl of Peterborough, have been so active that their progress looks altogether romantic, and will hardly be believed by posterity. They have not only maintained their conquest of the whole principality of Catalonia, but they have gained the kingdom of Valencia, and carried their arms as far as

and exhibits the valour, enterprise, and temerity of the English commander in strong contrast with the pusillanimity and credulity of the Spaniards.

Lieutenant-Colonel Cunningham, of "Cunningham's" Dragoons, having died of a wound on the 26th of January, 1706, Colonel Robert Killigrew, of the Royal Regiment of Dragoons, was appointed to succeed him in the command of that regiment.

The service of the Royal Dragoons at this period partook rather of the nature of guerilla warfare, and severely tested the discipline, courage, and intelligence of the men. Divided into small parties, and associated with bands of armed peasantry, they were continually making night searches among woods and mountains; hovering about the rear and flanks of the Spaniards, keeping them in constant alarm, such services being performed in concert with spies; and although under such circumstances it must have been difficult to preserve subordination and discipline, yet the regiment performed these duties to the satisfaction of the commander-in-chief. A vast tract of country was thus delivered from the enemy; and not the least peculiar incident of the campaign was, that Peterborough, being deficient in cavalry, procured 800 Spanish horses, and constituted Lord Barrymore's regiment of foot, now the 13th "Prince Albert's" Light Infantry, a corps of dragoons of which he appointed

Alicant; at the same time they blockaded Roses, though the two places were above four hundred miles one from the other."—*Present State of Europe*, January, 1706.

Lieutenant-Colonel Edward Pearce colonel. The regiment was equipped with accoutrements which had been ordered to be left at Vimaroz, and as cavalry it did good service throughout the subsequent campaign in Spain.

Lord Peterborough returned to Valencia on the 4th of February, 1706, where, amidst universal and enthusiastic rejoicings, he received a patent from the King constituting him viceroy of that kingdom.

The Archduke Charles had made his solemn entry into Barcelona on the 28th of the preceding October, where he had been again proclaimed King of Spain; but the unexpected surrender of that city, and the successes in Valencia having roused the Duke of Anjou and his grandfather Louis XIV. to renewed exertions, an attempt was determined upon for the recovery of Barcelona; and an army of upwards of 20,000 men, under Marshal Tessé, accompanied by Philip in person, entered Catalonia on the 8th of March, 1706; while about the same time a blockade was established by sea by a squadron under the Conte de Toulouse. In these circumstances the Earl of Peterborough with 800 horse, including the Royal Dragoons and 2,000 foot aided by a body of Miquelets, hastened from Valencia, and with this force he carried on an incessant guerilla warfare, keeping the French almost besieged within their own lines, which they had taken up in the beginning of April, on the 26th of which month they took by assault the fort of Montjuick, Lord Donegal, the commander, having been killed on the 10th previous, and the garrison forced to retire into the city.

While, however, the French batteries opened, and a general assault was daily expected to be made on the place, an English fleet with reinforcements on board under Lieutenant-General Stanhope, appeared off the port, when the French admiral immediately raised the blockade and put to sea—an example soon followed on shore by Marshal Tessé, who definitely raised the siege on the 12th of May and retreated towards Rousillon, leaving behind his artillery, ammunition, stores, and sick and wounded men. A squadron of the Royal Dragoons, with some other cavalry, were sent in pursuit of the retiring French, and being joined by hundreds of armed peasantry they fell upon their rear guard several times and took a number of prisoners. The Spaniards killed every man who fell into their hands, but the prisoners made by the English and Dutch were well treated.

After the flight of the enemy from before Barcelona the Royal Dragoons returned to Valencia, whence they expected to advance with King Charles upon Madrid, where the allied army, commanded by the Marquis das Minas and the Earl of Galway, had arrived towards the end of June. But irresolution, delay, and obstinacy on the part of Charles, the want of union among his generals, and the return into Spain of the French and Spanish forces after the raising of the siege of Barcelona uniting with the troops under the Duke of Berwick, compelled the allies to abandon the capital, and caused material and unfortunate changes in their operations.

In July the Royal Dragoons left Valencia, together with "Pearce's" newly-formed Dragoons, a regiment of Castilian foot and a regiment of Germans, and on the 8th of August they joined the army of Portugal at Guadalaxara, where on the 6th the Archduke Charles and the Earl of Peterborough had preceded them with reinforcements, thence marching to Chinchon, a town of Toledo, sixteen miles from Madrid, where they remained about a month.

At Guadalaxara the Earl of Peterborough, no longer on the most confidential or friendly terms with the Archduke Charles, and disgusted with the jealousies and vexations he found in the conduct of affairs, withdrew from the army with which his services had been so brilliant and so valuable, and left Spain for Italy.

The allied army, unable to make head against the superior numbers of the enemy, and being also in circumstances of much discouragement, broke up from their cantonments and commenced their retreat from Guadalaxara on the 28th of August, when the Royal Dragoons crossed the Tagus at Fuente Duennas, and continuing their march through the fine champaign country of La Mancha, took up their winter quarters at Valencia.

In the spring of 1707, being ordered to take the field, the regiment was detached on the 9th of April to Denia, and while encamped at Collera, a town at the mouth of the river Xucar, in Valencia, the battle of Almanza was fought on the 28th of April, when the allied army, commanded by the Marquis das Minas and the Earl of Galway, was nearly annihilated by the French and Spaniards under the Duke of Berwick. Soon after this disaster the Royal Dragoons joined the wreck of the allies which had been collected by the Earl of Galway, and were engaged for three months in marches and countermarches, observing the motions of the enemy and endeavouring to preserve the rich and extensive province of Catalonia from their power. They formed afterwards part of the force assembled for the relief of Lerida, but which was found to be impracticable, the kingdoms of Arragon and Valencia being occupied by the enemy. Catalonia was now the only portion of Spain remaining in the hands of Charles. The Earl of Galway soon after Almanza resigned the command of the British army in that country.

The journal of Lieutenant-Colonel de St. Pierre does not extend beyond this period, but from it sufficient extracts have been made, it is hoped, to establish its claim both to general and regimental interest. It may also in these days be amusing to read the following lists of the effects packed in the colonel's trunks, as showing the field equipment of a field-officer of cavalry in 1706. It may be noted that, with respect to uniform, no alteration appears to have been made since the warrant of 1684 :—

"Mémoire de ce que contient la grande caisse No. 1.

Une vieille couverture de cheval et caparaison vieux.
2 couvertures de cheval nuives pour coffres, &c.
Un caparaison neuf.
Une housse de cheval et les fourreaux de pistolets d'escarlate avec galon d'or et frange d'or.
Un surtout de camelot escarlate.
Une veste d'etoffe noire.
Une culette de drap escarlatte avec du galon d'argent.
Une veste et culotte de toile frise verdatre.
Une robe de chambre de Damas pouceaux et vert.
Un petit pacquet de treilles noir qui contient de quoy faire veste, culotte et doubleure grise.
Des boutons et du cordonet pour les habits de livrée.
3 paires de bas de toile grise verte.
2 paires de fuêtres de toile grise.
4 nappes de damaseè et 16 serviettes le tout marqué et numerotè.
10 callecons.
12 chemises.
12 coifes de nuiet marquées par nombre.
12 mouchoirs savoir six fines et six moins fins.
6 bonnets de nuit, dont trois sont de futaine ragèe, et trois autres de piqueurs de toile de coton.
2 paires de bas de laine tout neuf et 4 autres presque tout neuf.
Un habit de drap gris blanc une veste et culotte, boutons d'argent.
Un vieux surtout d'escarlate doublé de noir.
Un vieux surtout de drap bleu.
Un habit de drap noir doublé de taffetas avec veste et culotte.
Une veste d'etoffe noire.
Une vieille veste de toile grise avec une doubleure aussi de toile.
Une piece de drap de livrée.
Une paire de drap de tout neuf qui enveloppeur le linge et les meilleurs habits.
Un petit pacquet un valet nouveau.
4 camisoles de toile fine avec des manches.
4 sous manches.
2 camisoles de futaine.
Une roquelaure d'escarlate à boutons d'or.
2 paires de pistolets.

"Mémoire de ce qu'il y a dans la malle.

4 grosses chemises de nuit.
2 paires de draps tout neufs marqués et numerotés.
3 layes d'oréiller.

4 chemises assez fines.
6 autres assez grosses.
4 autres sans vien au poignet.
2 camisoles de futaine, et une autre de futaine sans manches.
2 camisoles de toile sans manches.
2 calecons neufs et 4 vieux.
10 coiffes de nuiet.
6 mouchoirs neufs et six vieux.
6 vieilles serviettes unies pour la barbe.
3 cravates de tarlatane une assez fine avec layes d'or.
Cravate a petites dentelles.
4 cravates de nuiet de mousseline close.
4 autres plus estroite de mesme mousseline.
10 autres cravates fines de differente largeur dont trois avec de la frange.

" Mémoire de ce qu'il y a dans la grande caisse No. 6.
Une grande tente de coutil et deux de Dragons.
2 paires de bottes à moy l'une neufves et l'autre non.
5 paires de souliers neufs et une paire de pantoufles.
Une selle de velours a troussequin et deux autres.
7 vieilles housses.
4 paires de culottes pour Dragoons et une veste.
2 selles et deux brides complettes.
3 paires de pistolets.
Plucieurs semturons cartouches, &c.
4 paires de bottes.
1 paire des cordes de fourrage.
Une caisse composée de chocolat et livres.
Six couteaux a manche d'argent.
6 cuilères et six fourchettes d'argent.
1 paire de petits flambeaux, un gobelet d'argent et salière.
Une housse bleue et chaperon ; du passier tous les articles sont dans la petite caisse.
Dans un des paniers il y a un gros surtout doublé de noir.
Un ruste corps gris doublé de gris.
Une veste de drap noir.
Une autre veste de soige noire.
Une culotte de drap bleu vieille.
Une petite bride verte.
Une platine de mousquet et crochets pour les armes."

The latest entry in the journal of the strength of the regiment is of the 17th of May, 1706, when three troops, amounting to about 135 men, were under the command of Lieutenant-Colonel de St. Pierre at Cuillera. The colonel died in the year 1713.

On taking the field in 1708 the Royal Dragoons were reported to be in "excellent condition,"[1] but in the campaign of this year their service was chiefly on outpost duties in Catalonia and Valencia, the allied army being now commanded by Marshal Count Guido de Staremberg, an officer of reputation, who had commanded the Imperial troops in Hungary.

The regiment wintered in Catalonia. In the campaign of the following year it was similarly employed in defensive operations, encamping for a considerable time on the banks of the Segré, when, in the month of August, 1709, the towns of Balaguer and Ager were captured and garrisons placed in them.

The campaign of 1710 was illustrated by more important events, the two claimants to the Spanish throne leading their armies in person. The Duke of Anjou commenced operations by the siege of Balaguer, but on the approach of the allies he retired; and when the Archduke Charles joined his army, the Royal Dragoons were detached from their camp on the Segré to meet and escort him to the camp.

After some manœuvring, Lieutenant-General (afterwards Earl) Stanhope, who commanded the British troops in Spain, being at the head of the leading column of the allies on the march towards Alfaras, discovered,

[1] *The Present State of Europe for* 1708.

on the morning of the 27th of July, a body of the enemy in front of the village of Almanara, in Catalonia, and obtained permission from the Archduke to attack them with the cavalry, of which the Royal Regiment of Dragoons had the honour to form part on this memorable occasion.

The sun was going down on the horizon, and the shades of evening were deepening over the valleys of Catalonia, when the British commander led forward his warlike horsemen. Before him appeared twenty-two squadrons of Castilian cavalry, the pride and flower of the Spanish army, with Philip's Life Guards on the right; a second line of the same strength was seen in rear, and nine battalions of Infantry supported the Cavalry. Against this force the gallant Stanhope advanced at the head of Harvey's Horse, now 2nd Dragoon Guards, "Queen's Bays." His front line consisted of sixteen squadrons with a reserve of six squadrons. The Spaniards came on with all the pride of war, when the opposing lines dashing into each other at full speed, the contest was of short duration. The left of the enemy soon gave way; the Life Guards were routed, with the loss of a standard and a pair of kettledrums; their second line fled in confusion; the supporting infantry were seized with panic, and Stanhope's brave troopers chased the fugitives from the field with great slaughter, following them up among the rocks and hills until darkness rendered it no longer possible to distinguish friends from foes.

The result of this brilliant action of cavalry disconcerted the projects of Philip, who, calling in his detachments, retired; the allies following up the pursuit for many days and making themselves masters of several towns in Aragon, until, on the 18th of August, the Royal Dragoons overtaking the rear-guard in the Pass of Penalva a sharp skirmish ensued, in which Lieutenant-Colonel Colberg, who commanded the regiment, was wounded and taken prisoner.

Continuing the pursuit, the Royal Dragoons crossed the Ebro with the leading column under Major-General Carpenter; and on the evening of the 19th of August the French and Spanish forces appeared in order of battle to the right of Saragossa, a large and rich city lying on the river Ebro, in a fine country. Preparations were immediately made for an attack on the following day, the Royals forming part of the cavalry of the left wing commanded by Lieutenant-General Stanhope, and opposed to the right of the enemy posted on the brow of a steep hill.

Early on the morning of the 20th of August a heavy cannonade commenced; and as the mountains re-echoed the sound, while the smoke, tinged with the rays of the sun, rose in clouds and formed a sparkling dome over the opposing hosts, the Archduke Charles and his suite galloped along the line, his presence infusing a glowing ardour among the troops. About mid-day Lieutenant-General Stanhope led the Royal Dragoons and other British horse against their adversaries, when, in the course of a severe contest, the superior numbers of the French had the advantage; but Stanhope's second line of cavalry repulsed the enemy; and the British dragoons rallying and returning to the charge, a sanguinary conflict took place at the foot of the hill. Six squadrons of Portuguese dragoons on the extreme left fled without waiting for the attack of the troops advancing against them. The battle extended along the front to the banks of the Ebro, and the Imperial, Dutch, and Palatine troops vied with the British in feats of gallantry. The Royals, Peffer's, now 8th R. I. Hussars, and Stanhope's dragoons gained some advantage. Harvey's Horse also signalised themselves; and four English battalions commanded by Major-General Wade being mixed with the cavalry of the left wing, behaved with remarkable heroism and intrepidity. Throwing off their knapsacks, they sprang up the acclivity and attacked their opponents sword in hand. Finally, the enemy were driven from the field with prodigious slaughter and the loss of 6,000 prisoners, twenty-two pieces of artillery, seventy-two standards and colours, the ammunition, baggage, and the plate of Philip; and, to complete the tale, the city of Saragossa with its stores, ammunition, provisions, and clothing became the prize of the victors in this memorable engagement. The Royal Dragoons passed the night in the fields near the city, and were thanked by Charles for their distinguished gallantry.

After the victory the allies again advanced to Madrid, where Charles made his public entry on the

28th of September, but the army of Portugal not moving to support this operation, the most disastrous results were the consequence. The Duke of Anjou called to his aid troops from Estramadura; reinforcements reached him from France, the Castilian peasantry took arms in his behalf, and once more the allies were forced to retire.

On the 11th of November the Archduke Charles withdrew from the army, taking with him the Royal Regiment of Dragoons and Staremberg's Imperialists, and proceeded to Cienpoznelos, and thence to Barcelona, escorted by two squadrons of the Royals. The third squadron remained with the army, and during the retreat it formed part of the rear column commanded by Lieutenant-General Stanhope, which retrogade movement was performed under great difficulties, owing to the hostility of the Castilians, inclement weather, and a scarcity of forage and provisions. On the 6th of December this column arrived at Brihuega, a town of about 1,000 houses, situate in the mountains of Castile, near the river Tajuna, where it halted the following day, and while here the place was suddenly surrounded by the French and Spanish forces under the Duc de Vendôme, the newly-appointed commander-in-chief. The British, though without artillery, with very little ammunition, and invested by a force of more than ten times their own number, made a vigorous defence; but the enemy forced the gates, battered down part of the walls, and after two unsuccessful attempts to storm the town, the British eventually were compelled to surrender, and to the number of more than 2,000 men became prisoners of war.

The English troops thus made prisoners at Brihuega on the 9th of December, 1710, were as follows :—

Harvey's Horse, now 2nd Dragoon Guards.
Royal Dragoons, 1 Squadron.
Pepper's Dragoons, now 6th Hussars.
Stanhope's Dragoons, disbanded.
Foot Guards, 1 Battalion.
Harrison's Foot, now the 6th.
Wade's „ „ 33rd.
Dormer's „ disbanded.
Bowle's „ „
Gore's „ „
Munden's „ „
Dalzel's „ „

Nearly at the same time the Dutch under Staremberg were defeated at Villavisioza, and with this unfortunate episode ended virtually the "War of Succession" in Spain, throughout whose long and trying campaigns the Royal Regiment of Dragoons had never failed signally and universally to uphold the reputation of the British cavalry.

The officers and men of the Royal Dragoons taken at Brihuega were sent to France, and after being exchanged were removed to England, and subsequently to Scotland. The remainder of the regiment continued to serve in Spain under the Duke of Argyll.

In 1711 the Emperor Joseph died, the Archduke Charles left Spain for Germany, where he was elected

Emperor of the Romans, which thereby removed one of the competitors for the throne of Spain. The Duke of Anjou made a formal renunciation of his claim to succeed to the crown of France, and a general pacification of Europe was arrived at by the Peace of Utrecht on the 11th of April, 1713.

In the summer of 1712 the officers and men of the Royal Dragoons quitted Spain, and having sold the Spanish horses upon which they had been mounted, they returned to England dismounted.

CHAPTER III.

HOME SERVICE.

In the reign of Queen Anne scarlet was definitively established as the uniform of the British Army.

After their return to England the regiment was dispersed in various quarters in Yorkshire, and the establishment was fixed at twenty-seven officers, six quartermasters, and 326 officers and men. In the summer of 1713 a detachment proceeded to Dover, there to receive a draft of 200 horses from "Ker's" Dragoons, that corps being ordered to Ireland dismounted, and there to be disbanded.

Queen Anne died on the 1st of August, 1714, when the Royal Dragoons left Yorkshire for the neighbourhood of London; but after the arrival of King George I. from Hanover they returned to the North, when a reduction of eighty men was made in the etablishment.[1]

By a Royal warrant of 3rd of February, 1715, addressed to Colonel William Ker, it was ordered that

[1] Marching Order Books and Establishment Books in the War Office.

two troops of the Royal Dragoons, three of the Scots Greys, and one newly-raised troop of Dragoons, should be formed into a regiment and reconstructed as "Ker's" Dragoons, which had been disbanded in 1713, retaining its former rank and standing in the Army. It was at the same time styled "The Prince of Wales's Own Royal Regiment." The two junior troops of the Royal Dragoons thus transferred were commanded by Captains Lewis Dalton and Peter Renouard, and the regiment thus raised has become the present 7th, or Queen's Own Hussars.

Colonel the Honourable William Ker, third son of Robert, third Earl of Roxburgh, died a Lieutenant-General in the army on the 7th of January, 1741.

The establishment of the regiment was thus reduced to six troops, and on the 13th of June this year the colonelcy was conferred upon Richard, Lord Cobham.

At this period Jacobite principles were very prevalent in the United Kingdom, and in September, 1715, the Earl of Mar raised the standard of rebellion in Scotland, and excited the clans to take arms in favour of the Pretender, James Stuart. The Royal Dragoons were immediately ordered to the North, and reaching Edinburgh in the early part of October, they went, under the orders of Lieutenant-General Carpenter, in pursuit of the rebels. After several marches and countermarches, Carpenter arrived at Jedburgh on the 30th of October, where, finding that a division of the rebel army had marched in the direction of Carlisle, he instantly started in pursuit of them. The rebels, however, reached Preston, in Lancashire, without opposition, where also arrived, on Sunday, the 18th of November, Lieutenant-General Carpenter, with the Royals, Molesworth's, and Churchill's Dragoons, two newly-raised corps, afterwards disbanded, and here they found the town surrounded by the troops of Major-General Wills. Some sharp fighting had already taken place, but before the arrival of the force from Scotland the town had surrendered. On the same day another division of the rebels under the Earl of Mar was defeated at Sheriffmuir, near Dumblane, by the Duke of Argyll. On the 22nd of December James Stuart landed at Peterhead with a suite of six officers only, and found his affairs in a condition so hopeless that on the 4th of February, 1716, he embarked with Mar at Montrose, and returned to France, the insurrection both in Scotland and England being completely suppressed.

In the year 1716 regiments were first numbered, having hitherto been distinguished by the names of their colonels. The Royal Dragoons, however, were never otherwise designated.

After the suppression of the rebellion the regiment was stationed in Leicestershire and Nottinghamshire, whence, in February, 1717, they moved to Newcastle-on-Tyne, and came under the command of Major-General Wills. This change appears to have been occasioned by the preparations made by Charles XII. of Sweden for supporting renewed pretensions of James

Stuart to the British crown, but which were rendered useless by the precautions of the Government and by the death of Charles. The journals of this period speak highly of the condition of the British Army, particularly of the cavalry, which they represent as the "best in the world."[1]

In the spring of 1718 the regiment went into quarters in Yorkshire and Lancashire, the establishment being reduced to 207 officers and men.

The peace of Europe was menaced by Philip V. of Spain in 1719, who, desiring to recover the places ceded by him in the treaty of Utrecht, among other measures contemplated was the placing of the Pretender, James Stuart, on the British throne, in order that the favourable interest of this country might be thus secured. An expedition was prepared under the Duke of Ormond for a descent upon the coast of England, but the fleet was dispersed by a storm. Two ships, however, having on board the Marquis of Tullibardine and the Earls Marischal and Seaforth, reached Scotland, where, on the 10th of April, these landed at Kintail, in Ross-shire, with about 300 Spaniards, who were joined by some hundreds of Highlanders. Intelligence of this event reaching London, orders were despatched for the Royal Dragoons to proceed with all possible speed to Scotland, where they arrived in May. On the 10th of June Major-General Wrightman, with a body of foot and three troops of the

[1] *Annals of George I.*, &c.

Scots Greys, attacked the Spaniards and the Highlanders at the Pass of Glenshiel, forcing them to retire with considerable loss, and on the following day the Highlanders dispersed and the Spaniards surrendered themselves prisoners of war.

The Royal Dragoons returned to England in July to be quartered in Yorkshire, while a detachment was ordered to Portsmouth, there to embark with an expedition commanded by their colonel, Viscount Cobham, and intended for an attack upon Corunna. The design upon that place was, however, abandoned, but the troops effected a landing on the coast of Spain, and took Vigo, where they captured seven pieces of brass ordnance, with a magazine of muskets and other arms.

Rondendella and Pont-a-Vedra also were taken, and additional seizures made of military stores. The Spanish Court made overtures for peace, and in November the expedition returned to England.

In February, 1720, his Majesty issued a regulation fixing the amount of purchase-money to be paid for regimental commissions, and the following prices were established for the Royal Regiment of Dragoons :—

Colonel and Captain	£7,000
Lieutenant-Colonel and Captain	3,200
Major and Captain	2,600
Captain	1,800
Captain-Lieutenant	1,000
Lieutenant	800
Cornet	600
Adjutant	200

The Lieutenant of the Colonel's troop was styled Captain-Lieutenant.

The Royal Dragoons left Yorkshire in April, 1721, for Nottingham and Derby, and on the 19th of that month the colonelcy was conferred upon Sir Charles Hotham, Bart., Viscount Cobham being removed to the 2nd Horse, now the 1st or King's Dragoon Guards.

During the summer of 1722 the regiment was encamped near Durham; and on the 12th of January, 1723, the colonelcy, having become vacant by the death of Sir Charles Hotham, was conferred upon Brigadier-General Humphrey Gore, from the 10th Dragoons, the present Prince of Wales's Own Royal Hussars.

The regiment was stationed in Nottinghamshire and Derbyshire in 1724, and in the following year it furnished a detachment to assist the revenue officers in their duties on the coast. In October, 1726, it was stationed in Sussex and Essex. In this year the Royal Dragoons were augmented to nine troops of 852 officers and men, and were selected to form part of the force of 10,000 men to be furnished by England in aid of the States-General in their war with the Emperor of Germany, but no embarkation was required.

The demise of King George I. took place on the 10th of June, 1727, and a few days previous to the coronation of his successor, George II., on the 10th of October, the Royal Dragoons marched into quarters near London, and were reviewed by His Majesty in brigade with Honeywood's Dragoons, now the 11th or Prince Albert's Hussars, on Hounslow Heath, on the 17th of the same month. They subsequently moved into Leicestershire and Derbyshire, and in the beginning of 1728 the establishment was again reduced to six troops.

In the spring of 1730 the regiment removed into Worcestershire and Gloucestershire. In 1731 it was stationed in Kent with detachments on coast duty, and in the month of March of the following year proceeded into Somersetshire, where, in the spring of 1733, it detached several parties to the towns and villages on the Suffolk coast, where frequent encounters took place between the military and the smugglers.

The various detachments being collected in May, 1734, and the six troops assembled at Taunton in Somersetshire, they were reviewed by their colonel, Major-General Gore. One troop was afterwards detached into Sussex, and in August another troop proceeded to Bath, and furnished a daily guard to the Princess Amelia during the residence of her Royal Highness in that city. In August, 1738, the five troops in Somersetshire marched to the North, there to be under the command of Lieutenant-General Wade, commanding the forces in Scotland, but in April, 1737, they returned to England to be quartered in Lancashire, and during the following summer the six troops were stationed in Essex and Kent with detachments on coast duty.

In July, 1739, the Royal Dragoons were ordered to call in their detachments and to take up quarters at Hounslow and its vicinity, where, on the 28th of that month, they were reviewed by his Majesty. In the beginning of August they moved into Worcestershire, and Major-General Gore dying on the 10th of the month, his Majesty bestowed the colonelcy of the regiment upon Charles, second Duke of Marlborough, from the 33rd Foot.

In this year, the Spaniards having repeatedly violated the existing treaties in regard to the trade of England with America, King George II. declared war against Spain, and the establishment of the Royal Regiment of Dragoons was augmented to 438 officers and men.

In May, 1740, the colonelcy, vacant by the removal of the Duke of Marlborough to the second troop, now second regiment, of Life Guards, was conferred on the 12th of that month upon Major-General Hawley, from the 13th Dragoons, the present 13th Hussars.

During the summer the regiment was encamped with three other regiments of cavalry and six of infantry near Newbury, and afterwards near Devizes, under the orders of General Wade. In October it went into Leicestershire.

In November, 1741, the Royal Dragoons moved into Somersetshire, and when in the summer of 1742 King George II. sent 16,000 men into Flanders under Field-Marshal the Earl of Stair, for the purpose of assisting Austria against France, Bavaria, and Prussia, the regiment was selected for this service, and after being reviewed by his Majesty on Hounslow Heath, they embarked in August, and on arrival in Flanders they were quartered in the cavalry barracks at Ghent.

Leaving Ghent in February, 1743, the regiment

marched into Germany, and in June it encamped with the other forces near Aschaffenburg on the river Maine, where they were joined by the King and the Duke of Cumberland. On the 26th of June the army marched on Hanau, a town of Hesse-Cassel, the Royal Dragoons forming part of the advance column, and while on the march the French army commanded by the Duc de Noailles, showed itself in position near the village of Dettingen in Bavaria. His Majesty immediately made his dispositions for attack, the Royal Dragoons, under Lieutenant-Colonel Naizon, being near the right of the line. Lieutenant-General Hawley, colonel of the regiment in conjunction with Lieutenant-General Cope, commanded the second line of horse during the battle. *London Gazette*, 16th July, 1743.

The French advancing to attack the left of the allies, the action soon became general, and the English cavalry encountered the cuirassiers with varied success.

The mousquetaires noirs, a *corps d'élite* of French cavalry, separating themselves from their line, and passing between two columns of infantry, dashed headlong upon the British horse, but the Royal Dragoons undauntedly met them in mid-career, overthrew their squadrons, cut them down with terrible effect, and captured a standard taken by a sergeant of the right squadron. It was of white satin embroidered with gold and silver ; in the centre a sheaf of nine arrows tied with a wreath, with the motto *Alterius Jovis altera tela*. The lance was broken, the standard stained with blood, and the cornet carrying it was killed without falling, being buckled to his horse, and the standard buckled to him.[1] The regiment was afterwards engaged with the French Household troops, and although without cuirasses was again victorious over its steel-clad opponents, and received the thanks of his Majesty, himself a witness of their spirited conduct. Eventually, the French army was overthrown, and fled from the field with great loss.

In this battle the Royal Dragoons had six men and thirty-four horses killed and ten wounded, and the regiment has been authorised to bear the word "Dettingen" on its guidon in commemoration of its services on that occasion.

Having passed the night in the fields adjacent to the field of battle, exposed to a heavy storm of rain, the Royal Dragoons on the following days marched to, and encamped on, the banks of the river Kinzig, remaining there until in the early part of August they advanced, and having crossed the Rhine above Mentz, they were employed in operations in West Germany. Nothing of importance, however, occurred, and in October they began their march to Mentz, there repassing the Rhine, and continuing through the Duchy of Nassau, the principality of Liége, and the province of Brabant, they entered Flanders, and reaching Ghent on the 10th of November, they again occupied part of the cavalry barracks there.

The campaign of 1744 passed without any general

[1] *London Gazette.*

engagement, and the services of the Royal Dragoons were limited to pickets, outguards, and protecting foraging parties from the attacks of the French, and in October they returned once more to Ghent.

In April, 1745, the regiment left their winter quarters, and encamped near Brussels. The enemy assembled in force and invested Tournay, the chief town in the province of Hainault, when the Duke of Cumberland, though vastly inferior in numbers, resolved to attack them. His Royal Highness advanced, and on the 10th of May a squadron of the Royal Dragoons was engaged with other troops in driving in the French pickets and outguards. Their army of 76,000 men, commanded by Marshal Saxe, appeared in order of battle, formed on a gentle ascent, and protected by batteries rising gradually from the plains near Fontenoy, a Belgian village in the province of Hainault, and at daybreak on the 10th of May the allies moved forward, but having much difficult ground to traverse and ascend, the attack did not commence until ten o'clock. The British and Hanoverian infantry pressed forward, and throughout the day displayed the greatest valour, but the Dutch by no means showed equal resolution, and their failure occasioned the most disastrous results. It was near the conclusion of the action before the Royal Dragoons were called upon to charge, when they advanced, led by Lieutenant-Colonel Naizon through a hollow way full of obstacles, and were exposed to a destructive fire from two batteries. They charged by alternate squadrons with all the spirit and determination which characterise the attack of British cavalry; but the Duke of Cumberland, perceiving that from the failure of the Dutch, and other causes, it was impossible to retrieve the fortunes of the day, decided upon a retreat, which was conducted in perfect order as far as the town of Aeth, near which the army encamped.

The loss of the regiment in this engagement was fifteen men and sixty-nine horses killed, with Lieutenant-Colonel Naizon, Cornets Hartwell, Desmeret, and Creighton, thirty-one men, and forty-seven horses wounded.

The allied army afterwards encamped on the plain of the Dender, near Lessines, and subsequently near Brussels.

Meanwhile events of consequence were taking place at home, where, on the 28th of July, Charles Edward, son of the Pretender, James Stuart, landed in Scotland with the determination of making a desperate attempt to seize the crown, and judging the moment favourable owing to the King's troops being so much employed on the Continent. Several regiments were immediately recalled to England, and among them the Royal Dragoons, who in the month of November marched to Williamstadt, and there embarked; the sailing of the ship, however, was delayed some time by contrary winds, and several horses were lost by the stranding of the transports.

Upon arrival in England the regiment formed part

of the army assembled near London for the purpose of repelling a threatened descent of a French force upon the south-eastern coast of the kingdom.

The rebellion in the north having been suppressed by the victory at Culloden, on the 16th of April, 1746, the Royals continued in the south at Windsor, Reading, and Colmbrook, and had the honour of furnishing travelling escorts for the Royal family. In July one troop attended Princess Caroline to Bath. On the 26th of December, 1747, the regiment was reviewed by King George II. on Hounslow Heath, and during the ensuing summer of 1748 it was employed on coast duty in Lincolnshire, and in the suppression of riots among the weavers in Lancashire. After the peace of Aix-la-Chapelle, in October this year, the establishment was reduced to 285 officers and men, and in 1750 the regiment moved to Scotland.

In 1751 a regulation was issued relative to the clothing and standards of the several regiments, from which the following particulars have been extracted relative to the Royal Dragoons :—

" *Coats*—Scarlet, double-breasted, without lappels, lined with blue; the button holes worked with narrow yellow lace; the buttons of yellow metal, set on two and two; a long slash pocket in each skirt; and a yellow worsted aiguillette on the right shoulder.

" *Waistcoats and Breeches*—Blue.

" *Hats*—Bound with gold lace, and ornamented with a yellow metal loop, and a black cockade.

"*Boots*—Of jacked leather.

"*Cloaks*—Of scarlet cloth with a blue collar, and lined with blue shalloon; the buttons set on two and two upon yellow frogs or loops, with a blue stripe down the centre.

"*Horse Furniture*—Of scarlet cloth; the holster caps and housings having a border of royal lace with a blue stripe down the centre; the crest of England within the garter embroidered on each corner of the housing, and on the holster caps the King's cypher and crown with I.D. underneath.

"*Officers*—Distinguished by gold lace; their coats and waistcoats bound with gold embroidery; the button holes worked with gold; and a crimson silk sash worn across the left shoulder.

"*Quarter-masters*—To wear a crimson sash round the waist.

"*Sergeants*—To have narrow gold lace on the cuffs, pockets, and shoulder straps; gold shoulder knots or aiguillettes, and yellow and blue worsted sashes tied round the waist.

"*Drummers and Hautboys*—Clothed in scarlet coats lined with blue, and ornamented with royal lace with a blue stripe down the centre; their waistcoats and breeches of blue cloth.

"*Guidons*—The first, or King's guidon to be of crimson silk, embroidered and fringed with gold and silver; in the centre the rose and thistle conjoined and crown over them, with the motto *Dieu et mon droit* underneath; the white horse in a compartment in the first and fourth corners, and I.D. in gold characters on a blue ground in a compartment in the second and third corners. The second and third guidons to be of blue silk; in the centre the crest of England within the garter on a crimson ground; the white horse on a scarlet ground in the first and fourth compartments, and I.D. within a wreath of roses and thistles upon a scarlet ground in the second and third compartments."

By the above warrant a special arrangement of the loops of lace on the coat in a trefoil pattern was sanctioned for the Royal Regiment of Dragoons, and for no other corps.

In 1752 the Royal Dragoons returned to England, and were stationed at York, whence they marched in October, 1753, into Norfolk and Essex, and in September the following year they moved into Kent.

Disputes having arisen between England and France relating to the boundaries of the British possessions in North America in 1755, an augmentation of 100 men was made to the establishment, and a light troop consisting of three officers, one quarter-master, two sergeants, three corporals, two drummers, and sixty privates,[1] was raised and added to the regiment on the principle of the light companies to regiments of infantry.

War was declared against France in 1756, when, the French making preparations for a descent upon the British coast, the Royal Dragoons occupied the maritime

[1] War Office Establishment Book.

towns in the southern counties until, in the summer of 1757, they encamped near Salisbury.

The military strength of Great Britain having been considerably augmented, his Majesty prepared to act offensively, and in 1758 the light troop of the Royal Dragoons formed part of an expedition commanded by Charles, Duke of Marlborough, which, landing on the coast of Brittany, destroyed the shipping and magazines at St. Malo. This troop was afterwards engaged in a second expedition under General Bligh, when a landing being effected in the Baie des Marées, Cherbourg was taken and put under contribution, when a force under the Duc d'Harcourt drove them out, but not until much damage had been done to the town.

On the 5th of April, 1759, the colonelcy of the regiment, having become vacant by the death of General Hawley, was conferred upon Lieutenant-General the Honourable Henry Seymour Conway, from the 4th Irish Horse, now the 7th Dragoon Guards; and in this same year the establishment of the six heavy troops was increased to sixty privates, and the light troops to eighty-nine, making a total of 544 officers and men; and in the year following the light troop was further augmented to four officers, one quarter-master, four sergeants, four corporals, two drummers, and 100 privates.

In the meantime an army commanded by the Marquis of Granby had proceeded to Germany, and was there serving in conjunction with the Hanoverian, Hessian, and Brunswick troops, commanded by Prince Ferdinand, Duke of Brunswick; and in the spring of 1760 the Royal Regiment of Dragoons, under Lieutenant-Colonel James Johnston, embarked for foreign service, and landing at Bremen, in Lower Saxony, on the 16th and 17th of April, they joined the army encamped near Fritzlar, in the principality of Lower Hesse, on the 21st of the month. On the day following they were reviewed by the Duke of Brunswick, who was pleased to express his approbation of their appearance.

After much manœuvring and skirmishing, 30,000 French troops, commanded by the Chevalier de Muy, crossed the Diemel, with the intention of cutting off the communication of the allied army with Westphalia. The Royal Dragoons with several other corps were immediately sent forward to Liebenau, under the orders of the Hereditary Prince Charles of Brunswick, and being followed by the main body, his Highness advanced to the vicinity of Warburg, and reconnoitred the French positions there with the intention of attacking them the next day.

At daybreak on the 31st of July the Royal Dragoons, under Lieutenant-Colonel Johnston, left their camp on the heights of Corbach, and making a detour gained the left flank of the French army, and several other corps arriving together at the same point, an attack was immediately commenced. After a severe contest the enemy gave way and retired upon Warburg, where he was again attacked and driven across the Diemel with great loss. The Royal Dragoons encountered the cavalry corps of the Royal Piedmont, and acquitted themselves with their accustomed gallantry. They afterwards charged a corps of Swiss infantry, the regiment of Planta, with great bravery, broke its ranks, and after killing many of the men took prisoners 20 officers and 200 soldiers; many of the Swiss attempting to escape were drowned in the Diemel. Three troops of the regiment followed the Marquis of Granby in pursuit of the enemy across the Diemel and halted for the night on the heights of Wilda; the other three, having suffered severely in the charge on the Swiss infantry, remained at Warburg.[1] In a general order issued on the occasion, Prince Ferdinand declared that "all the British cavalry performed prodigies of valour."

The Royal Dragoons lost in this action 3 men and 21 horses killed, and 13 men and 13 horses wounded.

The regiment subsequently encamped on the banks of the Diemel, and on the 18th of October it was despatched towards the Lower Rhine, forming a separate corps under the Hereditary Prince Charles, which invested Wesel, a town situated in the Duchy of Cleves.

The enemy advanced in force to relieve this place, and encamped on the 14th of October behind the convent of Campen. On the evening of the same day, after dark, the Royal Dragoons and other corps advanced with the intention of surprising the enemy in the night, but it

[1] Journal of Lieutenant-Colonel Johnston, Royal Dragoons, MS.

being found necessary to dislodge a corps which occupied the convent, the firing which this occasioned gave the alarm, and the troops immediately formed in order of battle. The action commenced on the following morning before daybreak, and a succession of attacks, repulses, and charges were kept up until nine at night, in which the Royal Dragoons took an active part, and they are reported to have "behaved extremely well." Two pieces of cannon and a pair of colours were captured; but at length the Prince perceived that it was impossible to drive the enemy out of a wood, of which he had possessed himself, and the allied regiments having expended all their ammunition, his Highness ordered a retreat.

In this affair the casualties of the regiment were heavy, comprising 8 men and 10 horses killed; Lieutenant-Colonel Johnston, 3 men, and 4 horses wounded; Captain Wilson and Lieutenant Goldsworthy, Cornet Duffe and 25 men taken prisoners. It afterwards repassed the Rhine on the 18th of October, and was cantoned in the Principality of Hesse, where the officers received orders to wear mourning for His Majesty King George II., whose demise had occurred on the 28th of the month.

In February, 1761, the regiment was engaged in an incursion into the French cantonments, and in several skirmishes with the enemy, and in the spring a remount joined from England.

The allied army, after much manœuvring, took up a position in Western Westphalia on the rivers Asse and

Lippe, the Royal Dragoons encamping on the heights between Illingen and Hohenover.[1] On the 15th of July the French attacked the troops under the Marquis of Granby at Kirchdenkern, when, to relieve them, the Royals crossed the Asse by the bridge at Hans Hohenover, to support the corps engaged. After a sharp action the enemy was repulsed with loss; but the firing of the skirmishers was maintained through the night, and on the following morning the enemy renewed the engagement with great fury. The regiment was posted near Vellinghausen, and when the attacking columns were repulsed it advanced to charge, but was prevented by the hedges and marshy hollows which intersected the country. It was subsequently employed in operations on the Diemel, and moving into the Electorate of Hanover, it was engaged in a skirmish near Eimbeck, in the early part of November. On the same night, marching through heavy snow to Foorwohle, it encountered and drove back some French cavalry; and on the 9th of the month it had another skirmish at Foorwohle, after which it went into quarters in East Friesland.

Leaving their winter quarters in May, 1762, the Royal Dragoons, on the 18th of June, joined the army encamped at Brackel, in the bishopric of Paderborn, whence they marched to the heights of Tissel. The French, commanded by the Marshals d'Estrées and

[1] *Journal of the Campaigns in Germany.* By an Officer present with the Army.

Soubise, took post at Groebenstien, where, on the 24th of June, Prince Ferdinand, resolving to attack them, his army was pushed forward for that purpose in several columns.

Leaving their camp at daybreak, the Royal Dragoons crossed the Diemel at Liebenau at about four in the morning, and advanced upon the enemy's camp with such address that the troops were in presence of the French before these had the least apprehension of an attack, and being simultaneously assailed in front, flank, and rear, they retired in confusion, leaving all their equipage behind them. The Royals had advanced against the enemy in front, and they were afterwards employed in surrounding a division of the French army commanded by General Stainville in the woods of Wilhelmstadt, where several corps were made prisoners. The pursuit being continued, the French took refuge under the cannon of Cassel, when the regiment, retiring a few miles, encamped near Holtzhausen.

During the remainder of the campaign the regiment was employed in operations on the Fulda, the Eder, and the Lahn, which were of such success that a considerable portion of territory was wrested from the enemy, the allies also taking the city of Cassel. These advantages were followed by a treaty of peace, when the Royal Dragoons went into quarters in the bishopric of Münster.

At this moment Colonel James Johnston, who had commanded the regiment since the 7th of April, 1759, and during the campaign of 1762 had commanded a brigade composed of the Royals and 2nd Dragoon Guards, Queen's Bays, received a most flattering tribute of the approbation of the Hereditary Prince of Brunswick, afterwards reigning Duke, who married the Princess Augusta, sister of George III., and who died of wounds received at the battle of Jena, on the 10th of November, 1806, in the shape of a valuable gold snuff-box embellished with military trophies, and accompanied by an autograph letter of which the following is a copy:—

"MUNDEN, *le* 17 *de Nov.*, 1762.

"MONSIEUR,

"Vous m'obligerez sensiblement en acceptant la babiole que je joins ici, comme une marque de l'estime et de la considération parfaite que je vous porte, et comme un souvenir d'un ami qui jamais ne finira d'être,

"Monsieur,

"Votre très-humble et très-dévoué serviteur,

"CHARLES, PRINCE HÉRÉDITAIRE DE B.

"A MONSIEUR LE COLONEL JOHNSTON."

Colonel Johnston rose to the rank of General, and was at different periods Colonel of the 9th Light Dragoons, of the 1st Horse on the Irish establishment, the present 4th R.I. Dragoon Guards, and of the 6th Inniskillen Dragoons. He was also Governor of Quebec, and dying on the 13th of December, 1797, he was interred in Westminster Abbey. The General wrote a journal of the campaign in Germany in 1760, which was obligingly forwarded to the compiler of the Record of the Royal Regiment of Dragoons by his grandson, Major Johnston unattached.

In the course of the winter of 1762, 63 ships arrived from England to convey the troops home; and the Royal Dragoons, commencing their march in February to Wilhelmstadt, there embarked, the strength of the regiment, according to the official statistics, being 14 officers, 329 men, and 423 horses, with 24 servants and 38 women.

After their return from Germany, the regiment was ordered to Scotland; at the same time the light troop was disbanded and the establishment reduced to 231 officers and men. Eight men per troop were equipped as light dragoons and mounted upon small horses for skirmishing and other light services, the remainder being mounted upon large horses of superior height and power.

In 1764 the regiment moved to England, and an order was received to mount with long-tailed horses. On the 9th of May, Lieutenant-General the Honourable N. Seymour Conway having, for political reasons, resigned all his military appointments, the colonelcy of the Royal Dragoons was bestowed upon Major-General the Earl of Pembroke, who had recently distinguished himself in Germany.

The 6 drummers on the establishment were, in 1766, ordered to be replaced by trumpeters, and on the 4th of May in the following year King George III. reviewed the regiment in Hyde Park, and expressed his approval of their appearance and high state of discipline.

After the review they went to the north of England, and on the 19th of December this year a Royal warrant appeared regulating the clothing, horse furniture, and standards of the regiments of cavalry, which contained similar directions to those of the 18th of July, 1681.

In 1769 the regiment was stationed in Scotland; but in the year following it returned to England, and after occupying various stations in the southern and western counties during the years 1770–72, it was again reviewed by His Majesty on the 17th of May, 1773, on Finchley Common, when, according to the journals of that period, its excellent condition and correct manœuvring produced the approbation of the King, Princes, general officers, and other spectators.

In the course of the summer the Royals again proceeded to the north, and making a short stay in Yorkshire, they continued on to Scotland, passing there the summer of 1774; but returning south in the succeeding year, the regiment, on the 24th of May, 1777, was reviewed in Brigade with the Queen's Bays, on Wimbledon Common, by the King, accompanied by several of the young Princes, and attended by a retinue of noblemen and general officers.

In 1778, hostilities having commenced between Great Britain and the colonies in North America, an augmentation was made to the army, when 6 sergeants, 6 corporals, and 126 privates were added to the strength of the Royal Dragoons, which, with several other corps were encamped on Coxheath, near Maidstone, and there reviewed by the King.

In 1779 the soldiers of the regiment equipped as light dragoons, the light troops of the 3rd Dragoon Guards, the 6th and 11th Dragoons, were incorporated into a regiment, which was numbered the 20th Light Dragoons,[1] and during the summer the 3rd Dragoon Guards, the Royals, the 18th, 20th, and 21st Light Dragoons were encamped on Lexden Heath near Colchester.

During the great riots in London in the beginning of June, 1780, known as the "Gordon Riots," the Royals were ordered thither, and in the following year they went to Scotland, when, at the termination of the American War in 1783, the establishment was reduced to 238 officers and men.

The regiment left Scotland in 1784, and during the ensuing six years it occupied various quarters in the northern and western counties of England. On the breaking out of the revolutionary troubles in France the establishment was increased by 9 men per troop, and in the spring of 1790 the regiment marched to Scotland, returning south the year afterwards, and was employed in the repression of disturbances in Birmingham.

A further augmentation was made to the strength of the regiment in 1792, and again in the spring of 1793,

[1] Official Records, Adjutant-General's Office.

when 4 troops were ordered to be held in immediate readiness for foreign service.

The enormities committed by the French Republicans occasioned the war of coalition against the French Convention, which had in fact been declared on the 10th of February, 1791, and a British force being sent to assist the Dutch in Holland, on the 10th of June, 1793, the troops of the Royal Regiment of Dragoons, which, in the absence of a field officer, seem to have been commanded by Captain William Spencer, embarked for the Netherlands to join the army commanded by H.R.H. the Duke of York, K.G.

Landing at Ostend, these 4 troops marched up the country and made part of a force which drove a body of French from the Camp de Cæsar behind the Scheldt, on the 8th of August. They were with the covering army during the siege of Dunkirk, and when the attempt on that place was abandoned they were employed in operations on the frontiers of Flanders, where they had a sharp encounter with a corps of French cavalry on the 27th of October.

On the 28th of January, 1794, the colonelcy of the regiment, vacant by the death of the Earl of Pembroke, was conferred upon Major-General Philip Goldsworthy.

In the month of April, the 4 troops assembled with the army near Cateau, and were engaged in the general attack upon the enemy's positions at Prémont, on the 17th, when Captain-Lieutenant the Honourable Thomas Carlton of the Royals was killed. The siege of Landrécies was immediately commenced, the regiment forming part of the covering army; also on the 24th of April it took part in the affair at Villiers en Couche, in which the French lost 1200 men and three pieces of artillery, and when the 15th Light Dragoons, now 15th King's Hussars, so particularly distinguished themselves. The casualties in the Royal Dragoons were 1 man and 2 horses killed; 2 men and 3 horses wounded.

On the 20th of the month the regiment again distinguished itself at Cateau, where the enemy, having marched out of Cambray, attacked the British army at daybreak. The Duke of York detached the Royals with seven other regiments of cavalry to turn the left flank of the French, a movement which was attended with the most brilliant results, and the enemy overthrown with great slaughter: the rout became general; cavalry and infantry mingled in promiscuous crowds were scattered over the plain and fell beneath the sabres of the British dragoons, who captured the French commander, Lieutenant-General Chapny, and 35 pieces of cannon. The Duke of York, in his account of the action, observes, "The behaviour of the British cavalry has been beyond all praise." The Royal Dragoons were among the corps which were declared to have "acquired immortal honour." They lost upon this occasion 6 men and 12 horses killed, with Lieutenant Froom, 2 sergeants, 11 men, and 14 horses wounded.

The fall of Landrécies took place on the same day, the 26th of April, when the regiment marched to the vicinity of Tournay, where on the 10th of May they were again in action, but lost only 2 horses killed, and 1 man and 3 horses wounded. His Royal Highness reported that the troops "had well supported the reputation acquired on the 26th of last month."

In the attack upon the French positions on the 17th of May the regiment was in reserve, after which the army resumed its post before Tournay, where on the 22nd it was attacked by General Pichegru with a large force, on which Brown, in his journal, observes:—"A column of five or six thousand men made its appearance towards our left, on which account the brigade of Guards and the British heavy cavalry remained ready for action on their camp ground; but the French observing our advantageous situation, and dreading the thought of meeting the British cavalry a second time in the open plain, thought proper not to approach. Finally the enemy were repulsed at every point, and in the evening they retired."

At length the Austrians were defeated, and the enemy brought forward such preponderating numbers that no chance of success remaining to the British troops, the Duke of York decided upon a retreat, which was followed by the evacuation of Flanders.

In the meantime another squadron of the Royal Regiment of Dragoons had embarked in England for foreign service, but being driven back by severe weather, it was relanded, and in July this part of the regiment moved from Salisbury to Weymouth, in consequence of His

Majesty's visit to that place; and in October upon the King's return to London, it marched to Dorchester barracks.

During the winter the troops abroad were exposed to privations and hardships which occasioned the death of many men and horses. The weather was unusually severe. The Dutch people were favourable to the French, and the British troops in their retreat through Holland during hard frost and storms of snow and sleet were treated by the inhabitants as enemies; but arriving at length in the Duchy of Bremen they there found rest and hospitality.

The regiment was engaged in no further hostilities. During the summer of 1796 it was encamped on one of the plains in Westphalia, and in the ensuing winter embarking for England the 4 troops from the Continent joined the squadron at Dorchester in January, 1796. In July following the whole regiment encamped on Barham Downs, near Weymouth, brigaded with the Scots Greys and 3rd King's Own Dragoons, under Lieutenant-General Lord Cathcart, whence in September it moved into barracks at Canterbury.

" *Wanted, Volunteers, for His Majesty's* 1*st or Royal Regiment of Dragoons, commanded by*

MAJOR-GENERAL THE EARL OF PEMBROKE.

"All young men willing to serve in the above-named Regiment shall immediately enter into pay and good quarters by applying to the Commanding Officer at the Head-quarters in the City of Exeter, or at Axminster, or at St. Mary, Newton Bushell, or with a recruiting party stationed at Devizes, Wiltshire, when each volunteer shall receive His Majesty's full bounty of two guineas and half, with an addition of pay, and a crown to drink His Majesty's health, also a good horse, arms, cloaks, and accoutrements, with everything necessary to complete a gentleman Dragoon.

"Young men wishing to be entertained as Royal Dragoons must be well made, perfectly sound and healthy, having no bodily infirmity whatever, from the age of 16 to 21 years, and from 5 feet 8½ inches to 5 feet 11 inches high.

"No trampers nor vagabonds need apply, nor any seafaring men, and likewise militiamen not having served their time, or any apprentice whose indentures are not given up; nor will any man be entertained that is not known something of, as it is the intention of the Regiment to enlist none but honest fellows that wish to serve their King and country with honesty and fidelity.

"GOD SAVE THE KING!"

CHAPTER IV.

THE PENINSULA.

THE Royal Dragoons marched, in October, 1797, to Birmingham and Coventry; in July, 1798, to Exeter and Taunton, whence in the following summer they moved to Radipole barracks, Weymouth; and on the 10th of August the following order was received, in consequence of which the horses' tails were cut.

"GENERAL ORDERS.

"The heavy cavalry, with the exception of the two regiments of Life Guards and Royal Regiment of Horse Guards, are to be mounted on nag-tailed horses.

"The First, or King's Regiment of Dragoon Guards; the First, or Royal Regiment of Dragoons; the Third, or King's Own Regiment of Dragoons, are to be mounted on black nag-tailed horses.

"The Second, or Queen's Regiment of Dragoon Guards, are to be mounted on nag-tailed horses of the colours of bay and brown.

"The Second, or Royal North British Regiment of Dragoons, are to be mounted on nag-tailed grey horses.

"All other regiments of heavy cavalry on the British establishment are to be mounted on nag-tailed horses of the colours of bay, brown, and chestnut.

"The custom of mounting trumpeters on grey horses is to be discontinued, and they are in future to be mounted on horses of the colour or colours prescribed for the regiments to which they belong.

"HARRY CALVERT,
"*Adjutant-General.*

"HORSE GUARDS,
"10*th August*, 1799."

In November following the regiment moved to Salisbury.

In the course of the summer of the year 1800 an encampment of about 30,000 men was formed on Swinley Common, near Windsor, where the Royals arrived in July. The troops were frequently exercised in the presence of the Royal Family, and the King reviewed the several corps previous to their departure. On the 11th of August, the regiment quitted the camp for Croydon barracks and Epsom, with a squadron detached on coast duty in Sussex.

On the 7th of January, 1801, his Majesty conferred the colonelcy of the regiment upon Major-General Thomas Garth, from the 22nd Light Dragoons, in succession to Lieutenant-General Goldsworthy, deceased.

Towards the end of May the regiment moved to Canterbury, and furnished numerous detachments on revenue duty in the towns and villages on the coast of

Kent, where they assisted in making large seizures of smuggled goods, for which they received a reward of £1 per man.

A treaty of peace with the French Republic having been signed at Amiens, on the 27th of March, 1802, a reduction of 2 troops was made in the establishment, the officers being placed on half-pay. In July 4 troops were ordered to Trowbridge to assist the civil power in the suppression of riots, and in October the regiment moved to Exeter and Taunton, with detached troops on coast duty in Cornwall.

War with France was declared anew on the 10th of March, 1803, and in April the Royal Dragoons moved to Dorchester, Radipole, and Wareham barracks, whence in July to Arundel and Chichester, the establishment being at the same time increased from 8 to 10 troops.

In April, 1804, the regiment was stationed at Ipswich and Woodbridge, thence it proceeded in November to Colchester, where it passed the winter.

The Royal Dragoons left Colchester in April, 1805, for York, Newcastle-on-Tyne, and Birmingham. In January, 1806, they returned to Woodbridge, and in March ensuing they once more proceeded north, and arriving in Scotland the head-quarters were fixed at Edinburgh with detached troops at Dunbar, Haddington, and Perth, the regiment having marched upwards of 600 miles in the preceding three months.

Embarking from Scotland in January, 1807, the regiment proceeded to Ireland, from which country it had been absent one hundred and fifteen years, and on arrival the head-quarters were stationed at Dundalk, with detached troops at Belturbet, Lisburn, Monaghan, Sligo, Enniskillen, and Londonderry. In June, 1808, it marched to Dublin, with troops detached to Carlow and Athy.

The state of affairs in the Spanish peninsula now induced His Majesty's Government to send thither a force to the assistance of the patriots in Portugal and Spain, which two countries had been taken possession of by Napoleon, since the 18th of May, 1804, elected Emperor of the French. This force, under the command of Lieutenant-General the Honourable Sir Arthur Wellesley, K.B., landed in Portugal on the 1st of August, 1806, but after the victories of Rolica and Vimiera, those operations were brought to an end by the unsatisfactory Convention of Cintra, made with Marshal Junot on the 30th of the month by General Sir Hugh Dalrymple, Bart. The British being now commanded by Lieutenant-General Sir John Moore, K.B., advanced into Spain to the assistance of the patriots in that country upon whose throne had been placed by Napoleon his brother Joseph, and the Royal Regiment of Dragoons having been ordered to join that army arrived in Cork for embarkation for Lisbon ; but the news of the result of that expedition, and the battle of Corunna on the 16th of January, 1809, occasioned the order to be countermanded. The regiment in consequence remained at Cork until April, when it took up extensive cantonments with head-quarters at Clonmel, whence in August following it once more marched to Cork where 8 troops of 80 men and 80 horses each embarked for Portugal under Colonel the Honourable George de Grey. The transports sailed on the 2nd of September, and on the 12th and 13th of the month the regiment landed at Lisbon and occupied the barracks at Belem.

The condition of the Royal Dragoons on arrival in the Peninsula will be best understood by the following extract from a letter from Lieutenant-General Viscount Wellington, K.B., to Lieutenant - General Payne, commanding the cavalry:—

"LISBON, *Oct.* 10, 1809.

"MY DEAR GENERAL,

"I arrived here yesterday, and I saw the Royal Dragoons in the streets, and I think that in my life I have never seen a finer regiment. They are very strong, the horses in very good condition, and the regiment apparently in 'high order.'"

The British army in Portugal, commanded by Lord Wellington, was at this moment occupying quarters on the Mondego. The Royal Dragoons, in January, 1810, marched a few leagues up the country to Santarem and Torres Novas, in the province of Estremadura, whence in February they moved to Niza and Alphalo in the Alemtejo.

The enemy having a great superiority of numbers, the British general was reduced to the necessity of acting on the defensive, but not the less was he resolved to maintain as long as possible a frontier position; and when, in the end of April, Ciudad Rodrigo was threatened, the Royal Dragoons advanced to Belmonte, in the province of Beira, where they arrived on the 8th of May. The French, commanded by Marshal Massena, Prince of Essling, proved, however, to be so numerous that all hope of saving Ciudad Rodrigo was abandoned, and the Royals leaving Belmonte on the 9th of June, proceeded to Villa Velha, whence, on the 1st of July, they marched to Villa de Touro, and towards the end of the month to Alverca. Ciudad Rodrigo fell on the 10th of July, and on the 30th of the month Lieutenant-Colonel Wyndham of the regiment was taken prisoner while visiting the pickets there. Lieutenant-General Sir Stapleton Cotton, Bart., commanding the cavalry, recommended for the vacancy Major Clifton, of the 3rd Prince of Wales's Dragoon Guards, which was confirmed with the rank of Lieutenant-Colonel on the 22nd of November ensuing.

The advanced posts of the British army having removed to Frexadas, the French besieged and took Almeida on the 20th of August, and on the day following they attacked a squadron of the Royals and one of the 14th Light Dragoons on picket at Frexadas, under Major Dorville of the Royals. The enemy brought forward a superior force of cavalry supported by infantry, but the two squadrons, undaunted by the greater numbers, charged the French with signal gallantry and drove them from the field with the loss of many men killed and wounded and 8 prisoners. The Royals lost 2 men and 1 horse wounded.

In a despatch of the 28th August, writing of this affair, Lord Wellington observes :—"A picket of this regiment (Royals) made a gallant and successful charge on a party of the enemy's cavalry and infantry, and took some prisoners."

The allied army retiring, the Royal Dragoons were actively employed during the movement, and particularly in a skirmish at Alverca, on the 2nd of September, on the main road to Almeida, in which a sergeant was wounded. On the 19th of the month they reached Santa Combu Dao, and again, on the 21st, they had an affair in which 1 man was wounded and another wounded and taken prisoner.

Continuing, in connection with the 14th Light Dragoons, to cover the retreat to the heights of Busaco, the regiment was formed in reserve during the action of the 27th of September. During the continued retrograde movement to the celebrated lines of Torres Vedras, the regiment held its post in rear; and on the 8th of October near Pombal, the enemy pressing upon the line of march, a picket, led by Lieutenant Carden, charged gallantly, and drove them back with loss, but following up his advantage too far, the lieutenant and

1 man, both wounded, were taken prisoners. The picket, notwithstanding, captured and brought off a French cavalry officer. The enemy's leading corps, supported by heavy columns, still continued to harass the rear, and the temerity of their cavalry was again checked on the 9th near Quinta de Torre by a determined charge of a squadron of the Royals, which drove them back with loss behind a corps of infantry. This corps was too much for the squadron, which, having received a volley, withdrew, having lost 6 horses killed, 1 sergeant-major and 2 men wounded, and 4 men wounded and taken prisoners. On the following day, the British entered the fortified lines of Torres Vedras, which, after reconnoitring several times, the Prince of Essling declined to attack, and during the night of the 14th of November he retired. The day after a picket of the regiment sent in pursuit made prisoners a sergeant and 8 French dragoons.

The French now established themselves upon the heights of Santarem, the Royal Dragoons being stationed at Cuzalbiera, Guinta, St. Christol, and Porto de Mugem, whence they detached parties on picket and outpost duty.

Massena, having exhausted his resources and wasted the physical strength of his troops, retired from Santarem on the night of the 8th of March, 1811, a movement followed by the immediate advance of the whole of Lord Wellington's army, the cavalry taking the lead.

From this date frequent extracts are made from a

journal kept in the regiment from 1811 to 1816, which contains many details of a nature more intimate and accurate than are to be found in more general accounts and descriptions. This journal, now in the possession of the regiment, was presented to the officers in 1878 by Captain Green, 49th Regiment, whose father had been executor to Captain Sigismund Trafford of the Royal Dragoons, whose name is frequently mentioned in the pages of the journal, and who died in Paris in 1852.

At the commencement of this advance of the army the condition of the corps is mentioned as follows :—

"Previous to the advance, no regiment could be in more perfect order than were the Royals. The horses at that time were all black. Windsor had lately joined with an excellent remount of nearly eighty horses, which rendered each troop very effective.

"As the regiment was filing over Santarem bridge, Lord Wellington and his staff passed by and were particularly struck with the excellent condition of the horses, their coats so jet black and shining. In fact it is not too much to say, that there were not such grooms in the world as the Royals were in those days."

On the 7th of March the regiment had a skirmish with the enemy near Pecoloo, where they took 3 prisoners, and had 1 man and a horse wounded; on the next day they had also 1 man and a horse wounded, and on the 11th, near Pombal, they took prisoners 2 sergeants and 26 men.

Resuming the pursuit on the following day, the army came upon a body of French cavalry, artillery, and infantry posted on a high table-land near Redinha, where Lord Wellington ordered the troops to form line of battle, the Royal Dragoons being directed to support the attack of the infantry. Three cannon shots from the British centre were the signal to advance, when at once a magnificent scene of war presented itself. The woods seemed alive with troops, and in a few moments 30,000 men, in three lines, were moving across the plain in a gentle curve, while the cavalry and artillery, springing simultaneously from the centre and left wing, charged in the face of a general volley from the French batallions, which were instantly hidden in the smoke, and when that cleared away were no longer to be seen, having made a precipitate retreat to Condeixa.

Lord Wellington continuing his advance, the Royal Dragoons, on the 14th of March, supported a successful attack of the infantry upon a French force, in the mountains of Casal Nova, and again on the following day they supported the attack upon the enemy's position at Foz d'Aronce. On the 18th, near Sernadilla, they captured a sergeant, 12 men, and 12 mules, on which occasion they had only 1 man wounded.

On the 26th of the month, still hovering near the French army, a patrol of the regiment, commanded by Lieutenant Foster, with one of the 16th Light Dragoons

I 2

under Lieutenant Peisse, attacked a detachment of cavalry near Alverca with conspicuous bravery, killed several and made prisoners an officer and 37 men.

Of this affair a despatch from Viscount Wellington of the 27th of March, 1811, says:—

"I have received a report of a gallant action of our patrols yesterday evening, under the command of Lieutenant Peisse, of the 16th Light Dragoons, and Lieutenant Foster, of the Royals, who attacked a detachment of the enemy's cavalry between Alverca and Guarda, killed and wounded several of them, and took the officer and 37 men prisoners."

The regiment had one man wounded in an affair on the 28th of the month, when they captured an officer's baggage near Ardés.

On the 3rd of April, they were posted in reserve during the action at Sabugal, and following up the retreat of the French they captured some mules with baggage at Alfayates.

On the 7th the Royals were sent to the relief of a corps of Portuguese militia, commanded by Colonel Trant, which had taken post near Fort Conception, and within half a mile of a brigade of French infantry. The destruction of this militia had seemed inevitable, when suddenly two cannon shots were heard to the southward; the French formed squares to retire, when in a few minutes six squadrons of British cavalry, with a troop of horse artillery, came sweeping up the plain in their rear, and the Portuguese were saved. The enemy, however, contrived to effect their escape with the loss of about 300 men killed, wounded, and taken prisoners, with part of their baggage; the Royals took a drove of 14 bullocks and a horse.

The Royal Dragoons were now in excellent quarters at Barquilla, and had much need of the advantages they afforded, for since the advance commenced on the 1st of March, their splendid condition seems to have considerably fallen off, of which the chief causes were: "first, the excessive hardships and fatigues, roads almost impassable, long marches, cold and wet bivouacs; and, secondly, the lamentable deficiencies of the Commissary;" which told severely upon both men and horses. At Barquilla, however, vegetation in the month of April was so forward that half the horses were turned out to grass, and as the frontiers of Spain were as yet unpillaged by either army, the necessaries of life were easily procurable, and thus the month of April went off very well.

The allied army now blockading Almeida, the Prince of Essling advanced to relieve that place, and driving in the allied pickets on the 2nd of May, Major-General Slade's brigade retired behind Nava de Vater, their position with the army covering the blockade. On the 3rd, about 11 A.M., the regiment being drawn up in position, a remount of 50 men and horses was brought up by Cornet Trafford from Lisbon, but during the severe contest of this day in the village of Fuentes d'Onor, the regiment was not engaged. On the morning of the 5th, at half-past three, heavy firing began on the right, and at half-past four the pickets were driven in with considerable loss to the 38th regiment and Portuguese. The picket of the 16th Light Dragoons was also sharply attacked, and Cornet Belli wounded and taken prisoner. The Royals remained for four hours exposed to a heavy cannonade, which fortunately, however, produced little effect. Several brilliant charges were made during the day, in one of which Cornet Trafford distinguished himself, and the French Colonel Latour being knocked off his horse surrendered to Lieutenant Gubbins. Two squadrons of the Royals, under Lieutenant-Colonel Clifton, also made a most gallant charge upon the enemy's cavalry, took prisoners a sergeant and 24 men, and released a party of the Foot Guards, who had been taken by the French. A party of their cavalry made a splendid attack, and captured two guns of Captain Bull's troop of Horse Artillery, when a squadron of the Royal Dragoons, dashing forward, re-took the guns and brought them into the British lines, together with several prisoners.

It is related that in the course of the day, "Clarke had a most narrow escape upon his old horse 'Turk,' inasmuch as a shell knocked both rider and horse completely over, and burst a few yards distant. Indeed, Clarke's troop was the one that principally suffered."

The casualties of the Royal Dragoons in the battle of Fuentes d'Onor were 4 men killed and 36 wounded,

18 horses killed and 52 wounded. It was altogether a very trying day, after which they returned to Barquilla and Villa de Ceirva.

While in these quarters a draft of about 80 men and 100 horses joined with Captain Tomkins and Lieutenant Rose, the former by exchange from the Life Guards.

The attempt to relieve Almeida having failed, on the night of the 10th of May the French garrison in that place blew up the works, and rushing in one column out of the town, made their way through the blockading troops, and directed their march upon Villa de Ceirva, but finding there the Royal Dragoons, they changed their direction to Barba del Puerco. The detachment meanwhile at Villa de Ceirva turned out suddenly in the night, pursued and overtook the French column, which they attacked, and brought off a sergeant, 9 men, and some baggage, having two men wounded in the affair. The 4th and 34th Foot continued to follow up the enemy, who, however, made good their retreat.

Lord Wellington now went into Estremadura, but the Royals remained with the forces left on the frontiers of Portugal, near Ciudad Rodrigo, and were stationed to cover the front from Villa de Egua to Espejo.

The French army, reinforced, and now placed under the command of Marshal Marmont, Duc de Raguse, advancing on the morning of the 6th of June in two columns, the light division had been ordered to retire the previous evening from Gallegos upon Nave d'Aver, and thence upon Alfayates. The Royal Dragoons, under

Lieutenant-Colonel Clifton, and one troop of the 14th Light Dragoons, assembled at 3 A.M. at Gallegos, for the purpose of covering this retreat, and it must be noted that in withdrawing the picket, a man of the name of Banks deserted to the enemy, this being the only instance of so great a crime in the regiment that had ever happened, with one previous case of a soldier who was caught in the attempt, tried by court-martial, and shot.

At seven o'clock the enemy showed themselves to the extent of about 2,000 cavalry, 6,000 infantry, and 10 guns; but this overwhelming force was met by the British cavalry with a resolution and ability rarely equalled. The celebrated French cavalry General, Montbrun, in vain endeavoured to outflank the Royals and 14th, his squadrons were twice attacked and defeated, and the retreat of the Light Division was effected with little loss. The French lost a colonel, "shot by a man of the name of Burnside in a very masterly style." The regiment had 1 troop serjeant-major, 3 men, and 6 horses killed, and 9 men wounded. It is stated that "since the regiment left Lisbon it never turned out stronger or looked better than on the present occasion. The horses were perfectly recovered, and the large remount of Tomkins had brought the regiment nearly to its original strength."

Lieutenant-General Sir Brent Spencer, commanding in the absence of Viscount Wellington in Estremadura, thus reports to his lordship on these events:—

"It is with great pleasure I have to mention the very

admirable conduct of the Royals under the command of Lieutenant-Colonel Clifton, and one troop of the 14th Light Dragoons, which being all that were employed in covering the front from Villa de Egua to Espejo, were assembled at Gallegos, and retreated from thence agreeably to my directions, and notwithstanding all the efforts of General Montbrun, who commanded the French cavalry, to outflank the British, pressing them at the same time in front with eight pieces of cannon, their retreat to Nave d'Aver merits the highest commendation."

Major-General Slade speaks in much praise of Major Dorville of the Royal Dragoons, and of Captain Purvis of the same regiment, "who had opportunities of distinguishing themselves." On the 4th of June, Colonel the Honourable George de Grey had been promoted to the rank of Major-General, and placed in command of a brigade, composed of the 3rd "King's Own" and 4th "Queen's Own" Dragoons.

On the night of the 6th of June, Major-General Slade's brigade bivouacked at Alfayates, and crossing the Coa the following day went into bivouac in rear of the infantry. On the 11th the regiment moved to Atalaya, on the 15th to Castello Branco, and 16th to Ladoyera, where a supply of hats and overalls for the men arrived, of which they stood in much need, as may be judged by an account of their appearance at this time, viz. "An old pair of stockings, with one rusty spur at the heel, an old pair of shag breeches, and a hat indescribable" having been worn through "three long successive years." Returning for a day only to Castello Branco, the regiment marching by Villa Velha, Niza, Apulion, and Portalegre, arrived about the 23rd of June in the immediate vicinity of the town of Arronches in Alantejo, standing on a rising ground near the small river Caza, where in an extensive wood of cork-trees and gum cistas, the Royals remained for six weeks in huts constructed by the men, but the horses suffered considerably from exposure to the broiling sun, the cork-trees affording but little shade.

Early in July much inconvenience began to be felt; the heat was most oppressive, the noxious smells from the camp became a serious evil, and by some accident a fire broke out which did considerable mischief, all which circumstances occasioned a desirable change of position nearer to the river and in a purer atmosphere. Major-General Slade had his head-quarters in Arronches.

The outpost duties while at this camp of Arronches were extremely severe. They were supplied by the heavy brigade and the 9th, 11th, and 12th Light Dragoons. The 11th had but recently arrived from England, and their very first picket proved singularly unfortunate, the squadron being surprised, and 4 officers and 40 men taken prisoners before any alarm was given. So harassing was the duty that for five weeks the saddles of the 12th Light Dragoons were never off the horses' backs, and the regiment in consequence became so unfit that it was sent to the rear to a village called

Aslinear, and for a short time was attached to Slade's brigade.

On the morning of the 25th of July the Royal Dragoons marched for Portalegre, on the 26th to Apulion, on the 27th to Niza, and on the 28th they crossed the Tagus to Castello Branco on the 29th. Moving by separate squadrons, the regiment reassembled on the 3rd of August at Penamara, situated on an eminence overlooking perhaps the most fertile and beautiful valley in Portugal. Following the course of the Aqueda, the regiment again separated on the 9th of August, one squadron attending Lord Wellington in a reconnaissance in the vicinity of Ciudad Rodrigo, the blockade of which place being raised by the advance of Marshal Marmont, the head-quarters of Slade's brigade and the Royal Dragoons were established at Soita, the regiment occupying the surrounding villages lying among woods of chestnut-trees bounded by the river Coa.

While at Soita, Major Jervoise, who had long been in bad health, was at length obliged to go to the rear, but was unable to travel further than the small village of Gabbion, 5 leagues from Abrantes, where he died.

In these quiet villages of Soita, Kiuta, Vitrenda, Villa Toro, &c., lost in the foliage of the vast forests of chestnut-trees, the brigade remained five weeks, until, on the 23rd of September, it marched to Etenero, and there bivouacked outside the village, in which was Lieutenant-General Sir Stapleton Cotton and his staff.

On the 25th and 26th the Royals were drawn up in front of the position of Fuente Guinaldo, as were the whole of the British cavalry and artillery until, at 10 P.M., the brigade was ordered to fall back to their destined position at Agaberca, 4 leagues off. "At 4 in the morning of the 28th of September, the regiment being mounted and formed in square in a wood behind Agaberca, a deep grave was dug, and a corpse sewn up in a blanket was brought forth to be buried. At first it was thought that some private soldier had died, but what was the astonishment and regret of every officer to find that the dead man was Crosbie, who had gone to the rear only two days previously slightly indisposed, and whose death was most extraordinary. On the 24th of September he was affected by a slight dysentery occasioned by drinking new wine; he was advised by Langman to remain behind with the baggage, and to beware of taking cold. On the 28th he went with the hospital to Agaberca, and finding himself still much relaxed, sent for an infantry surgeon, who gave him 6 opium pills, with orders to take one per day, until he found the purging cease. Crosbie did not pay sufficient attention to the injunction of the surgeon, and ignorant of the composition of the pills, he took all the 6, each of which contained a sufficient quantity of opium for one dose. The effect was mortal. In an hour after he had taken the pills, he called his dragoon and told him to lay out his cloak and blanket and make his bed, as he felt sleepy; in the meantime he continued walking up and down the room exclaiming how happy he felt, and occasionally began singing. Alas! poor man! he was then under the effects of opium; his happiness was that which preceded death! He lay down on his bed, fell into a deep sleep, and slept for ever!"

Thus in the 25th year of his age died Captain Crosbie, a most amiable and promising young man, beloved and lamented by the whole regiment.

After the funeral ceremony, the Royals slowly retired through Puebla d'Azaon Nyon, Aldea de Ponte, and the enemy following up pretty closely, about twelve o'clock the skirmishers of both armies became engaged. About eight p.m. the action became warm and the 4th Division which defended Aldea de Ponte was hardly pressed. The brigades of Slade and de Grey which had been retiring by échelon of half squadrons were brought up. The village of Aldea de Ponte was occupied by the French, and Slade ordered Major Dorville of the Royals to make a charge with his squadron, which he did in a most dashing manner, carrying all before him, and coming out at the end of the village rejoined the brigade without the loss of a man. One man was wounded, and Lieutenant Ross and Cornet Trafford had their horses shot under them. The French, however, kept possession of the village, and the British army fell back upon its position at Soita where the brigades of Slade and de Grey bivouacked together.

While at Melho and Sentory, instructions were received for the incorporation of two troops with the depot, thus reducing the number in the Peninsula to six, the Royals having been for some time the only regiment with 8 troops in the field. The ineffectives and sick men of two troops were in consequence sent to England and the depot placed under Major Purvis.

On the 1st of November Major-General Slade's brigade, consisting of the 4th R.I. Dragoon Guards, the Royals and the 12th Light Dragoons, was ordered in advance to take its turn of duty at the outposts, and was cantoned in the villages of Ettiero, Gallegos, Espeha and El Bodon. The service here was light, one squadron of the Royals, with one of the 12th under Captain Phipps of the former was detached to El Bodon in attendance upon Major-General Craufurd who made occasional reconnaissances in the vicinity of Rodrigo. The great objection here was the want of straw for which the long withered grass was a very poor substitute besides filling the horses' stomachs with worms to a frightful extent. At this time Mr. Hawes the Commissary, who had been so long attached to the regiment, and of whom though in other respects a good sort of man, they had so much cause to complain, was succeeded by Deputy-Assistant-Commissary-General France.

On the 27th of November, their time of service at the outposts having expired, the brigade was relieved by a brigade of German Cavalry, and fell back to Sorero Peres, Slade's head-quarters being at Nevada, when, much to the astonishment of every one, instead of continuing their route to permanent quarters in the rear, the regiment was ordered to resume its post in front; the 12th were detained at Nevades, and on the 2nd of December, the Royals once more returned to Espeha. On the 6th the different brigades resumed their former stations, the Royals concentrating at Sorero Peres, whence on the 9th, they took up winter quarters at Neda and Travoca, head-quarters at the former, where in all respects they found excellent accommodation both for men and horses.

On the 9th of January, 1812, an order suddenly arrived for the brigade to advance, and Lord Wellington investing Ciudad Rodrigo, the Royal Dragoons took post at the village of Villa Crossina about two and a quarter miles from the city, where also Slade had his head-quarters. On the evening of the 19th, the cannonade from Rodrigo was heard to be particularly heavy, when about 10 P.M. it ceased, a certain indication that the place had fallen, and on the following morning the garrison marched out prisoners of war; the same day the weather began to break and settled into violent and continuous rain, which lasted throughout the whole of the month following.

At Villa Crossina a month's pay was issued, which, unhappily for the time, occasioned great disorder in the regiment, and one case of peculiar gravity must be noted, being that of a dragoon named Coutts, who attacked and ill-treated Sergeant Currie in so brutal a manner, that being tried by court-martial, he was sentenced to receive 900 lashes, of which 710 were inflicted, the remainder being reserved until he should have recovered; these last 190, however, were never inflicted.

About the last day in January, the Royals commenced their return to their old quarters, the last day's march being 3 leagues to Carassarda. Between Villa Turpina and Pinto runs the Coa, the fords of which river, in general but knee-deep, were now so swelled by excessive rains that it was necessary to send the baggage round by the bridge of Almeida, thus causing a delay of three days, during which the regiment was at Travalko and St. Jao de Presquere without it. This ford at the Coa proved indeed so deep that it was not passed without much difficulty. Two horses only could go abreast, and on the least deviation to right or left they instantly lost their footing. One private of the name of Acheson going two yards to his right was instantly swept away by the stream, and both man and horse perished.

Arriving at Travano and St. Jao de Presquere, the head-quarters were established at the latter, the roads during the last day's march having been almost impassible, and in crossing what in summer was but a small rivulet, Mr. Warner was washed off his horse and drowned.

St. Jao de Presquere is a large, ill-built and dirty village on the Douro, carrying on a considerable wine trade with Oporto, and here with an infamous market, the inhabitants ill-disposed, and in complete dulness, the Royals passed five weeks.

With the first week in March came orders for the march of the brigade into the Alemtejo, proceeding by Thomar to Abrantes, thence through Gabion, Crato, Aldea de Cabron, Cabela de Vidi, Estremos, and Villa Viciosa, all more or less good towns. This fine province of Alemtejo is almost wholly overrun with gum-cistas and immense forests of cork-trees.

From Villa Viciosa the Royals, crossing the Guadiana on the 16th of March, continued to Santa Martha in Spain, and on the day following to Villafranca, where they remained to the 19th, and thence to Ribuera with a squadron at Caniosa, the French occupying Avnandros, about 3 leagues distant. The brigade remained several days at Ribuera under the orders of Lieutenant-General Sir Thomas Graham, commanding in Spanish Estremadura.

From Villa Viciosa, Slade's brigade moved to Santa Martha, thence to Villafranca, and thence to Ribuera, whence after a few days it retired to Berlinga, a considerable place at the extremity of Estremadura Palia.

At this period the siege of Badajos was in progress, and some idea of the nature of the service required of the Royal Dragoons may be gathered from the circumstance that "Captain Kennedy Clark was ordered to proceed to Campilla, from thence to hover about the neighbourhood of Llerena della Creusa. He was then detached with 20 men from any part of the army at least 50 miles, was obliged to keep a constant look-out and never to enter a village except for an hour or two, and night and day constantly to change his position. For ten days and nights Clark continued in the performance of this painful duty, although the nights in Estremadura are extremely cold and dewy, in excessive contrast to the burning heat of the day; all added to an unhealthy climate, want of food, want of rest, and great anxiety."

Badajos fell on the 6th of April, when the Earl of Wellington, going into the north, left a force, including Major-General Slade's brigade, in Estremadura under Lieutenant-General Sir Rowland Hill.

The head-quarters of the French cavalry, commanded by General Soult, retired from Villafranca the day after the fall of Badajos, pursued by Lieutenant-General Sir Stapleton Cotton with Ponsonby's and Slade's brigades, the latter returning to Ribuera and Villafranca, at which place also were the head-quarters of Lieutenant-General Sir William Erskine, Bart., commanding the cavalry, and the outposts of Sir Rowland Hill's corps. The French cavalry opposed to them was commanded by General l'Allemann. The head-quarters of Sir Rowland Hill and of the infantry were in general at Almendralejo.

On the 14th of May occurred the brilliant affair of Almarez, and on the 17th the Royal Dragoons being at Ribuera, with a squadron at Usagre and a picket under

Cornet Trafford at Lleira, General l'Allemann, with four regiments of Dragoons, made a night advance, and drove in this picket with the loss of 4 or 5 men, when the regiment falling back to Villafranca, there joined the 3rd Dragoon Guards, the brigade spending the night of the 18th drawn up in the wood near the town. l'Allemann, who only wanted to make a reconnaissance, retired to Hermandres.

On the 20th of the month the regiment paraded for the purpose of witnessing the punishment of a corporal and 6 privates, who had on the 17th previous got drunk under circumstances peculiarly scandalous. Being on picket at Villa Garcia, during the night they had stopped and plundered a train of mules passing through with a supply of aquadente; and being tried by court-martial, they were sentenced, the corporal to reduction to the ranks and to receive 600 lashes, the privates to 400 lashes each, which were inflicted.

On the 28th, a squadron under Major Dorville posted at Lleira, ascertaining that the French cavalry under General l'Allemann were threatening his post, the Major, leaving a sergeant's picket in Lleira, with a support at a ford across a small brook in front of a wood, behind which he withdrew his squadron after dark, the French entire brigade at 3 in the morning entered Lleira, expecting there to surprise the squadron, but captured only the sergeant and his picket; and continuing their advance they came upon Major Dorville and his squadron formed up and ready to receive them, upon which the brigade retired, followed by the Royals for a certain distance, who then returned to their position.

On the 7th of June the advance-guard of Marshal Soult entered Llerena, and on the night of the 9th, Sir Rowland Hill moved his infantry from Almendralejo to Usagre.

At 4 o'clock on the morning of the 11th, Slade received orders to advance with his brigade, and to bivouac in a wood between the two small rivers that cross the road between Hinosa and Lleira, where he arrived accordingly about 10 A.M., unbridled, and began to cook, when in about an hour and a half a patrol came rapidly in and reported that the French cavalry had entered Lleira, and were advancing; which proved to be General l'Allemann, with the 17th and 27th Dragoons. On coming in sight of the British brigade already formed up, and ignorant of what might be in the wood in their rear, l'Allemann instantly went about and retreated pretty sharply, followed at a canter by Slade through Lanneia and Gourdas towards Maguilla, but not before the Royals in a petty affair of skirmishers had already lost a sergeant and several men. Slade, neglecting some more favourable opportunities, at length made a determined charge with 3 squadrons of the Royals, supported by the 3rd Dragoon Guards, than which nothing could be more successful, sabring many of the enemy and taking prisoner one of General l'Allemann's aides-de-camp, when suddenly there arose a cry of "Look to the right!" "Look to the right!"

where, indeed, there appeared a squadron which the French general had kept in reserve, but merely for the purpose of covering the retreat of his brigade; for be it observed that the French were rapidly making off to the rear, while the British, seized with unaccountable panic, had also turned about, so that for some minutes the extraordinary sight presented itself of two parties mutually running away from each other. At this moment, Captain Hutton of the Royals, who commanded a reserve squadron, made a most gallant charge, but in vain; he was totally unsupported, and the whole brigade in the greatest disorder, utterly regardless of all the exertions and appeals of their general and regimental officers, continued their disgraceful flight until victors and fugitives, equally overcome and exhausted by the overpowering heat and the clouds of thick dust, came to a standstill at Valencia, about 8 miles from Maguilla, where at length Slade was able to collect his regiments, and then retired to the wood behind Lleira, whence he addressed his report to Lieutenant-General Sir R. Hill of this most disastrous day, in which he says:—

"Nothing could exceed the gallantry displayed by the officers and men on this occasion. Sir Granby Calcraft and Lieutenant-Colonel Clifton particularly distinguished themselves, as well as all the officers present.

"I beg particularly to report the conduct of Brigade-Major Radclyffe, of the Royal Dragoons, to whom I feel particularly indebted for his assistance on this occasion."

The casualties of the Royal Dragoons in this deplorable affair were 1 sergeant, 11 men, 6 horses killed; 19 men, 8 horses wounded; and Lieutenant Windsor, who was badly wounded, and 4 sergeants and 39 men taken prisoners. The brigade generally sustained a loss of 140 killed and taken prisoners; 2 officers taken, and more than 200 horses killed or died of excessive fatigue.

The Earl of Wellington, as may be supposed, was very far from being satisfied with the results of the 11th of June, as will appear from the following extract from a letter of his Lordship to Lieutenant-General Sir Rowland Hill, K.C.B. :—

"SALAMANCA, 18*th June*, 1812.

"MY DEAR HILL,

"The Royals and the 3rd Dragoon Guards were the best regiments in the cavalry in this country, and it annoys me particularly that the misfortune should have occurred to them; I do not wonder at the French boasting of it. It is the greatest blow they have struck.

"WELLINGTON."

The 12th and 13th of June the brigade remained at Hinosa, whence, on the 14th, a detachment of 28 men, with an equal number of the 3rd Dragoon Guards, under the command of Lieutenant Strenowitz, aide-de-camp to Lieutenant-General Sir William Erskine, Bart., had an affair with a French squadron at Maguilla, of whom they killed many, and captured the officer commanding, a sergeant, and 20 men, with their horses.

The enemy occupying Maguilla and advancing, Sir Rowland Hill retired, and on the 21st of June established a bivouac in the forest of Albuhera, where for 3 weeks the troops experienced much distress from the extreme heat, the want of water, and the tainted atmosphere, which caused a great deal of sickness. Here the brigade was joined by the 4th R.I. Dragoon Guards under Colonel Sherlock.

Early in July the French retired, when Sir Rowland Hill re-occupied his positions at Almendralejo and Villafranca, where news came of the victory of Salamanca on the 22nd of the month. The brigade now moved to Fuente del Maestre.

On the 17th of August this year appeared a warrant by which the cocked hat and jacked-boots so long worn by the heavy cavalry were superseded by a helmet and grey cloth overalls.

The Earl of Wellington had entered Madrid on the 12th of August, and Sir Rowland Hill making a corresponding move in advance, the Royal Dragoons marched on the 27th of the month, and on the 6th of September reached Villanueva, leaving which place on the 13th, and crossing the Tagus by the pontoon bridge at Almarez on the 19th, they arrived on the 28th at Talavera de la Reina, where at Temdeca in New Castile and in the neighbourhood of Madrid the regiment remained nearly two months, and encamped for about three days in the environs of the capital itself.

The siege of Burgos having commenced, Hill took up a position on the Tagus, the Royals marching by the royal palace of Aranjuez to Morata. The enemy, however, concentrating a very superior force, compelled the raising of the siege, and Wellington retiring, Hill made a corresponding movement, when the Royals commenced their retreat on the 27th of October by Madrid and the Pass of the Guadarama Mountains, arriving on the 12th of November at Salamanca, on the 18th at Arguilla; and having on the 17th had an affair in which 4 men and 1 horse were wounded, they went into quarters on the 28th in the miserable village of Zelreira, 16 miles from Alcantara, the head-quarters of Slade's brigade.

Wretched as was the village of Zelreira, the forage proved to be so good and abundant as speedily to produce a most beneficial change in the condition of the horses, which had suffered so terribly in this disastrous retreat of 30 days from Burgos, throughout which the hardships of the allied army were intense, and indeed seem in general to have recalled the celebrated retreat to Corunna in 1809, the scenes related of disorder and suffering being almost equally painful.

During the whole of the past year it will have been observed that the services of the Royal regiment of Dragoons were chiefly confined to duties of outposts and pickets, which, although in themselves of primary importance and calling for constant activity and intelligence, are peculiarly harassing, and not of a nature to gratify ambition of distinction.

A few days previous to the 1st of January, 1813, the regiment moved into Alcantara, between which city and their late quarters at Zelreira runs the small stream of the Ellas, dividing Portugal from Spain, and, as it turned out, in crossing by a narrow bridge the Royal Dragoons had left Portugal for ever.

In Alcantara, a large, well-built city with good, well-paved streets, and situated upon the almost perpendicular rocks overhanging the Tagus, the Royals remained all January, February, and part of the month of March, enjoying their excellent quarters.

While here took place at Broses, the head-quarters of Lieutenant-General Sir William Erskine, who commanded the cavalry of this division of the army, the tragical death of that officer, who, taking a fever, threw himself out of his bedroom window in a fit of delirium, and injured himself so terribly as to die three days afterwards, in the 47th year of his age. His funeral, arranged by his aide-de-camp, Captain Eckersley, of the Royal Dragoons, was attended by Major-General Slade's brigade, and a handsome tablet with a Latin inscription was erected to his memory.

About the 16th of March the brigade received orders to move to Las Narvos, distant from Alcantara about 16 miles, and here, on the eve of St. Patrick's Day, an order came to Colonel Sir Granby Calcraft, temporarily in command of the 2nd division of cavalry, vacant by the death of Sir William Erskine, and in the absence of Major-Generals Slade and Long at Lisbon, for the immediate transfer of the horses of the 4th R.I. Dragoon Guards to the 3rd Dragoon Guards and the Royal Dragoons, which accordingly was carried into effect on St. Patrick's Day, and the 4th proceeded dismounted to Lisbon, there to embark for England. This naturally was a cruel blow to Colonel Sherlock and his regiment, for although since their arrival in the Peninsula they had been unfortunate, they were now in high order, and could little have expected this misfortune.

Shortly after this circumstance Major-General Slade rejoined, and at Casa de Milan he gave up the command of his brigade, being succeeded on the last day of April by Major-General Fane.

In the second week in May, Major-General Fane's brigade received orders to leave Casa de Milan, and to join Lieutenant-General Sir R. Hill, whom accordingly they fell in with on the second day's march at a small village on the Alagon, and arriving at Salamanca on the 26th of the month, they forded the Tormes above the city, and encountering a body of French infantry with a few cavalry commanded by General Villatte, who was leaving Salamanca in the direction of Alba de Tormes, the right squadron of the Royals, led by Lieutenant-Colonel Clifton, charged with great spirit, sabred a number of the enemy, and made prisoners 143 men and 4 tumbrils. The regiment had 5 horses killed,

10 men and 3 horses wounded. Major Purvis's charger also was killed under him.

After this affair the brigade bivouacked, and the following morning were reviewed by Lord Wellington, and moving on a couple of leagues bivouacked with the whole army near La Orbada until on the 6th of June it took up a position to the right of the fine old city of Valladolid, once the capital of Spain. On the 17th the army descended the beautiful valley of the Ebro, than which nothing can be imagined more lovely, fertile, and interspersed with pleasant villages. Making a long march on the 19th, and bivouacking in an immense fir wood, the whole army on the following day concentrated at a position between 3 and 4 leagues from the city of Vittoria, in which vicinity took place on the 21st what may be considered the crowning victory of the Peninsular War.

Major-General Fane's brigade on that morning was quite in rear of the army and did not reach its position in line until about 8 A.M. The allied troops then continued slowly to advance, the French giving way, until, about 2 o'clock and about 3 miles from Vittoria, they made a decided stand with two very powerful batteries, which opened fire upon the Portuguese infantry, who behaved remarkably well, and in rear of whom and behind a wood was Fane's brigade, which had only a trumpeter, Wright, wounded by a cannon-shot taking off his foot, of which he died 3 months later, and whose horse was killed. At 3 in the afternoon Lieutenant-

General Sir Thomas Graham having effectually turned the enemy's position, the battle was over, ending in total rout and disorder, and the loss of the whole of their artillery, material, stores, &c., and the immediate evacuation of the city. The ground had been quite unfavourable to the operation of cavalry, who, however, towards evening pushed on in pursuit, the 3rd Dragoon Guards of Fane's brigade making a charge accompanied by Lieutenant-Colonel the Marquis of Tweeddale, Assistant-Adjutant-General, whose horse was killed, and while on the ground his lordship was considerably maltreated by some runaway French artillery drivers, who left him for dead in a ditch, from which he was not extricated for some time.

The casualties of the Royals were only 1 man wounded, 2 horses killed, and 1 horse wounded. They were commanded on this occasion by Major Purvis.

On the night of the battle of Vittoria, the brigade moved on about 3 leagues on the road to Pampeluna and there bivouacked. On the 22nd came down a terrible and incessant rain. Fane established himself in a village close by, and here took place the sale of the prize horses, wines, and other articles; after which an equitable distribution was made of the profits, amounting to about 20 dollars each to the subalterns, 90 to the captains, and 11 to the privates. Besides this several thousand sheep and nearly 200 bullocks were captured.

On the 27th of the month the brigade marched for Pampeluna, and occupied some villages in the vicinity of that city, where in particular the forage was most abundant, consisting chiefly of green barley and vetches. The difference of climate here was very perceptible, colder in fact than in England. The blockade of this important place commenced on the 28th of June; Fane's brigade, with the infantry division of Lieutenant-General the Honourable Sir Lowry Cole, and a Spanish corps, forming the investing force, the rest of the army occupying the passes of the Pyrenees at Roncesvalles and Irun.

On the 18th of July the brigade was moved to Sanguesa and Sambera, two towns on the river Aragon, 28 miles from Pampeluna; but on the 25th it suddenly returned thither, arriving on the 26th, and during the three days' fighting in the Pyrenees it remained in position, being employed on the 29th in collecting the wounded, who lay about among the trees and shrubs with which the mountains were covered. The Royals remained ten days at Suesta, Fane's head-quarters being at the village of Saragetta, and on the 10th of August the regiment returned to Sanguesa and Sambera, the 3rd Dragoon Guards occupying the village of Nybao, about a league distant. On the 12th of September the regiment was ordered to Villafranca, a town lying in the richest plain of Navarre, where indeed nothing could exceed the fertility of the soil and the beauty of the country watered by the rivers Aragon and Argas. A squadron at Sanao, and one at Mareilla, close by, were the out-quarters, and the 4 months spent here in such excellent circumstances, and in the absence of all duty, may be truly considered the most enjoyable period of the whole Peninsular War. Here also arrived from England a draft of 4 officers and 60 horses.

On the 26th of February, 1814, died, to the universal regret of the regiment, Captain George Hutton, a brave and excellent officer, a good man and a perfect gentleman.

He was buried with every mark of respect on the 28th, just outside the village of Rendinas in the Pyrenees, and on the grave was placed a stone inscribed as follows :—

In memoriam
GEORGE HUTTON,
Militis fortissimi
De prima legione
Equitatus Britannici
cohortis,
qui obiit
26 Calendas Februares,
Anno Domini 1814,
Ætatis suæ 36,
Hic Tumulus
Sacrietur.

On the 2nd day in March the Royal Regiment of Dragoons commenced its advance into France, arriving on the 3rd at Campananova, the 4th at Olita, and the 8th at Pampeluna, where it passed the following day. They reached on the eighth day the small town of Tolosa in Biscay, the whole road to which, from Pampeluna, is one continued ascent through scenery the most picturesque and attractive, though now partly covered with snow. The road itself, completed by Charles IV. at prodigious expense, is cut out of the solid rock. On the 9th, proceeding to Irun, the regiment there received camp equipage, and on the 11th of the month, crossing the Bidassoa, the Royal Dragoons entered France, a proud moment for a British army which, since the loss of Calais in 1559, had been unknown in that country. Their first night, after an extremely wet march, was spent at St. Jean de Luz, whence on the 13th they advanced towards Bayonne, arriving on the 18th at Dax, whence the brigade moved on by Aire towards Tarbes, where it joined the army after a march the longest and the most painful perhaps the regiment had ever made, lasting from 2 in the morning to half-past 11 at night.

Between Tarbes and Toulouse the brigade coming up with the French at St. Gaudin, Major-General Fane made a charge with the 12th Light Dragoons, supported by the 3rd Dragoon Guards, the Royals being about a mile in the rear, which was well conducted, and resulted in the capture of about 80 men and horses. The regiment then was posted at the village of Portel, on the Garonne, and about three leagues from Toulouse, the British videttes being posted in a line parallel to those of the French and within a few yards of them. On the 23rd of March the British army invested Toulouse, but it was not until the 10th of April that the action which decided the surrender of the city took place, and during this interval the regiment had pretty severe though monotonous duty at their post of Portel, but at length every preparation being completed, the army crossed the Garonne below Grenade, 7 miles from Toulouse, on the 9th and 10th of April, Hill's division and Fane's brigade of cavalry remaining in their usual position, so that the part taken by them in the battle was not more active than on the days preceding, but on the morning of the 14th, Hill's Division entering Toulouse at 8 o'clock, Fane's brigade filed over the great bridge and marched straight through the city, amidst great apparent enthusiasm for the Bourbons and the British army. Marshal Soult retiring by Villefranche and Carcassonne to Castelnaudary, the brigade, without halting, continued to a small village on the Canal de Languedoc, which that night it was unable to cross. It must be noted that this day's march had been one of peculiar fatigue and suffering, owing to the extreme violence of the wind and the clouds of dust, which very much distressed the column. The following day the Royals passed the canal and quartered in a village near Villefranche, whither came to them the news of the abdication of the Emperor Napoleon on the 4th of April previous, and the restoration of the Bourbon dynasty. The regiment now went into quarters in the village of Gardouche, near Villefranche, with which may be said to have ended the military operations of the Peninsular War.

In Villefranche, to which town the regiment moved, and which they had to themselves, the Royals spent an extremely pleasant time, everything being abundant and good. On the 24th of May the head-quarters moved to Montguiscard, and there arrangements were made for the proposed march of the Royal Dragoons with the rest of the cavalry through France to Calais, there to embark for England. On the 1st of June the sick and ineffective men, 5 officers, and the heavy baggage were sent by canal to Toulouse, and thence to Bordeaux, there to embark for England, and 36 horses considered unfit having been disposed of at Toulouse, on the 3rd of the month began the long march to Calais, which, terminating on the 17th of July, the regiment embarked the day following, and landed at Dover on the 19th, after an arduous and exemplary service on the Continent of 4 years and 11 months. The Royal Dragoons landed of the following strength :—

	Troop.						Horses.	
	Officers.	Quarter-Masters.	Sergeants.	Trumpeters.	Rank and File.	Total.	Officers.	Troop.
Disembarked	19	2	30	5	334	390	59	316
Sick absent	1	...	...	...	3	4	...	...
	5	2	6	...	104	118	...	13
Absent with and without leave	3	...	...	...	...	3	...	...
Prisoners of war	...	...	1	1	12	14	...	...
Total	29	4	37	6	453	529	59	329

CHAPTER V.

WATERLOO.

On the 27th of July, 1814, the regiment marched to Richmond in Surrey, preparatory to being passed in review by the Commander-in-Chief, Field-Marshal his Royal Highness the Duke of York, K.G., after which, marching to Newbury, it was there joined by the depot commanded by Major Dorville, and proceeded to Bristol, Bath, and Trowbridge, but with the exception of the troop at the latter place, the whole regiment united at Bristol under the command of Major Dorville, till the 11th of August, on the 24th of which month a reduction of two troops was made in the establishment.

At Bristol, upon the resumption of the command by Lieutenant-Colonel Clifton in September, the mess and band of the regiment were reconstituted, the former under the presidency of Captain Kennedy Clark, the latter under that of Captain Heathcote.

In the first week in December the regiment moved to Exeter, and now was granted the permission to bear on the guidons and appointments the word "Peninsula" as an honourable mark of distinction for its services during those campaigns. Lieutenant-Colonel Clifton also received a medal with one clasp for the battles of Fuentes d'Onor and Vittoria, and Major Purvis a medal for Vittoria. It will be noticed that the days were still far distant from the later general distributions of decorations and medals.

On the 4th of January, 1815, a squadron proceeded to Truro, there being also a squadron at Taunton, and head-quarters and 4 troops at Exeter.

The prospect of a lasting peace soon vanished; and the return of Napoleon to France, the flight of Louis XVIII. from Paris, the restoration of the Empire, the armed coalition of the leading Powers of Europe, and the appearance of a British army in the Netherlands, again called the Royal Regiment of Dragoons to the field. About 9 o'clock on the evening of the 21st of April, an express arrived at Exeter, directing the immediate march of the regiment to Canterbury for embarkation at Dover and Ramsgate, and on the 24th the Royals were *en route* for that destination, an augmentation of 2 troops being ordered at the same time. On the 13th of May 2 squadrons under Captain Phipps embarked at Ramsgate, and landed the following day at Ostend. On the 14th, Captain Clark's squadron marched into Canterbury from Truro, where the two succeeding days were employed in forming the depot to be left under the command of Major Purvis, and in receiving a transfer of 100 horses from the 5th

Dragoon Guards stationed at Canterbury. The head-quarters and the remainder of the regiment embarking at Dover on the 16th, they landed at Ostend on the following morning under disagreeable circumstances; for, the horses being let down into the sea to swim ashore, almost every animal was seized with a severe cold, which it required some time to get over. The town of Ostend also was so crowded that the same evening the regiment was compelled to march into the country 7 miles to a village called Ghisted. The day following they marched to Bruges, and on the 19th, Captain Phipps with his two squadrons having already arrived at Ghent, the whole regiment concentrated in the villages round that city for more than a week, a rest very much needed by the horses.

At this moment Ghent was crowded to excess, being the temporary refuge of Louis XVIII. and his fugitive Court, but on the 27th the Royals left the neighbourhood, and proceeded to occupy two or three small villages near the town of Ninove, where their head-quarters were fixed. The Life Guards and Blues were also in the town, and the whole of the British cavalry were cantoned in the villages adjacent. The Royal Dragoons were now in brigade with the Scots Greys and the Inniskillen Dragoons under Major-General the Honourable Sir William Ponsonby, thus forming the 2nd or the "Union" brigade of Cavalry.

On the 29th of May took place near Grammont, in the fine meadows on the banks of the Dender, and for which was paid from £400 to £500, that splendid review of the Cavalry and Royal Horse-artillery by his Grace Field-Marshal the Duke of Wellington, K.G., when Lieutenant-General the Earl of Uxbridge received the Commander-in-Chief at the head of 48 squadrons of the finest cavalry in the world, and of 8 troops of magnificent horse artillery, drawn up in three lines, a troop of artillery on either flank, and two troops in front of the centre of the second line. Brilliant weather favoured this grand military spectacle, and at half-past 2 o'clock a salute of 21 guns announced the approach of the illustrious chief, who came attended by an immense and sparkling *cortége*, among whom was conspicuous Marshal Marmont, Duc de Ragusa, mounted on the same white Arab he had ridden at the battle of Salamanca. After an inspection of the troops by his Grace, and a "march past," the review ended at 4 o'clock, when each regiment filed off to its cantonments, and until the 15th of June nothing occurred of any moment, the officers generally amusing themselves in visiting at each others' quarters, and in sport, particularly horse-racing, in which the Earl of Uxbridge took especial interest. On the 15th, however, certain rumours went abroad of an advance of the French, and at 5 o'clock on the morning of the 16th a sudden order of march arriving; at 6 o'clock precisely, Major-General Sir William Ponsonby's brigade was in motion, and continued to the town of Grummove where a momentary halt was made; about a league beyond this place the road became bad and hilly, which rendered the line of march, now moreover impeded by several other brigades of cavalry, very slow and tedious, so that this long and toilsome march of about 56 miles did not come to an end until 11 o'clock at night, when the brigade reached the position of Les Quatres Bras, where a severe conflict had taken place with the French under Marshal Ney, when the Royal Dragoons bivouacked in a field of barley in rear of the scene of action.

The Prussian army of Marshal the Prince Blucher had been defeated by the Emperor Napoleon at Ligny, on the 16th, and forced to retire, when the Duke of Wellington on the 17th made a corresponding movement, and in this operation the Royal Dragoons with the rest of the cavalry were employed to cover the retreat of the infantry upon the position of Waterloo. This service was admirably performed and exhibited a most interesting spectacle of warfare, the effect of the same being heightened by the circumstance that about half-past 6 in the evening, a tremendous storm of thunder and rain burst over the opposing armies, and rendered the manœuvring of the cavalry among fields of high standing corn extremely difficult. One squadron of the Royals was thrown out to skirmish under Major Radclyffe, who writes in his Journal:—

"I was detached with my squadron to cover the brigade by skirmishing, and Major-General Sir William Ponsonby, and the brigade generally, were pleased to applaud the style in which we acquitted ourselves. It rained with greater violence than I ever witnessed before, which I found to my advantage when it was my turn to skirmish. The enemy had 2 squadrons of Chasseurs opposed to me, and as they could not overpower us by their fire, they huzzaed and endeavoured to excite each other on with cries of '*Vive l'Empereur!*' and once actually charged towards my skirmishers, but they stopped short."

Towards evening the brigade reached the position in front of Waterloo, where it halted, and again passed the night in the open fields without provisions of any kind and exposed to continued rain. The only casualties in the regiment on the 17th were 1 man killed by a cannon-shot, and 2 wounded.

On the morning of the 18th, the allied army formed in order of battle, and Major Radclyffe's journal, speaking of the Royal Regiment of Dragoons, says :—

"We found ourselves in our place in close column behind the second line of infantry, fetlock deep in mud; no baggage for the officers, and neither provision nor water for the men, though some stray cattle had been killed and eaten, and a small supply of spirits had a short time before been found on the road, so that we might be said to go 'coolly' into action, for every man was wet to the skin."

About 10 o'clock the French army was seen forming on the opposite heights, whence a cloud of skirmishers soon sprang forward; the fire of the artillery gradually opened, and about noon their columns of attack came sweeping up the valley between the rival positions in all the pomp and majesty of warlike display. A succession of attacks were made at various points, while Sir William Ponsonby's brigade remained quietly in contiguous close columns of regiments waiting the moment when their services would be actively required. At length, about half-past 1 o'clock the infantry corps of the Count d'Erlon of 20,000 men suddenly appeared on the ridge opposite, and dashing forward with such celerity that, scarcely seeming to traverse the intervening space, they rapidly ascended the allied position, dispersed at once a Belgian brigade with which they had come in contact; forced the artillerymen posted in rear of the double hedge and narrow road to abandon their guns; broke through parts of the supporting British infantry, and several thousand having passed La Haye Sainte, the French had actually crowned the allied position. At this critical moment Lieutenant-General the Earl of Uxbridge galloped up to this part of the field, and uttering a few words, the "Union" brigade deployed at once into line; advanced, and halting for a few moments, to allow the broken infantry to retire through the intervals of squadrons, the 3 noble regiments led by their gallant General dashed forward with terrific violence upon the mass of the enemy's battalions. In an instant the heads of their columns were broken and forced back, a general flight commenced, the firing ceased, and, the smoke clearing away, those imposing masses, a moment before so conspicuous and conquering, had either almost disappeared, or left only a dispersed rabble flying in all directions. Everywhere the Royals, Greys, and Enniskillens were to be seen trampling down and sabring the fugitives with uncontrollable power. The "Eagle" of the 105th regiment of the Line was captured by Captain Kennedy Clark of the Royal Dragoons, that of the 48th by Sergeant Charles Ewart of the Greys, and 2,000 prisoners were taken in this splendid charge. Unfortunately, all did not end with this brilliant success, for the brigade, encouraged and excited by their victory, followed up their advantage too far. They swept across the ravine, carried several batteries, and continued their wild career even to the rear of the enemy's position, who, recovering confidence from the disorder too apparent of the British Dragoons, now fell upon them with a large body of Lancers and some Cuirassiers. The brigade, broken and disorganised in the pursuit, was driven back with heavy loss, in which was unhappily to be included that of its gallant leader, Major-General the Honourable Sir William Ponsonby, whose horse being blown, and unable to extricate him from the heavy ground of a ploughed field, the General was set upon and killed by the Lancers. This charge obtained universal admiration, and the conduct of the regiments engaged has been much commended by historians. After returning from the charge, the Royals resumed their former position, and became exposed to a heavy cannonade. The brigade, now commanded by Colonel Muter[1] of the Inniskillens, moved later in the afternoon

[1] Now Lieutenant-General Sir Joseph Straton, K.C.H. and C.B., Colonel of the 8th Royal Irish Hussars, who was authorised to take and use the surname of *Straton*, instead of Muter, on the 28th of September, 1816.

to the right, when Colonel Muter being wounded, he was succeeded by Lieutenant-Colonel Clifton of the Royals, the command of the regiment devolving on Lieutenant-Colonel Dorville. The French continued to make the most desperate but fruitless attacks upon various points, until at length the Duke of Wellington assumed the offensive; the allied army made a simultaneous advance; the enemy gave way, were overthrown, cut down, and driven with dreadful slaughter from the field, and the Prussian army of Blucher coming up to continue the pursuit, ended a day glorious beyond all precedent for the British arms. In his Grace's despatch the conduct of the heavy cavalry throughout this great conflict was especially noticed.

The Royal Regiment of Dragoons on this memorable occasion suffered severely. Captain Windsor, Lieutenant Foster, Cornets Magniac and Sykes, Adjutant Shepley, 6 sergeants, 86 men, and 151 horses were killed; Brevet-Major Radclyffe, Captain Kennedy Clark, Lieutenants Gunning, Keily, Trafford, Wyndowe, Ommaney, Blois, and Goodenough, with 6 sergeants, 82 men, and 35 horses, wounded and taken prisoners.

On the following morning the allied army advanced towards Paris, and the Emperor Napoleon abdicating on the 22nd of the month, the capitulation of the capital followed on the 3rd of July. The brigade, including the Royals, went into quarters at Nanterre, a small town 7 miles from Paris, on the 7th of July, whence, on the 28th, it marched to Rouen, in which grand old city, and in pleasant circumstances, it remained, until, in the month of October, the quarters of the brigade were changed, the Royals going to Montvilliers, in the vicinity of Le Havre, the Greys to Harfleur, and the Inniskillens to Bolbec.

On the 3rd of December the whole of the British troops were in motion, preparatory to the return of a portion to England, and the arrangements for the army of occupation to remain in France, and on the 4th the regiment left Montvilliers and marched by Dieppe to Abbeville, where the head-quarters of the cavalry were established, and where, on the day after their arrival, the horses were drafted into the regiments destined to remain in the country. During the stay of the regiment at Abbeville, the cold was so intense that several men on duty perished during the night.

The Royals proceeded to the small hamlet of Hardingen, about 4 leagues from Calais. The embarkation of the cavalry went on slowly, until on the 1st of January, 1816, the regiment in their turn embarked, and on the following day landed at Dover and Ramsgate, proceeding thence to the barracks at Ipswich, where, on the 28th of the month, the establishment was reduced from 10 to 8 troops.

CHAPTER VI.

TAKING OF THE EAGLE BY CAPTAIN KENNEDY CLARK AT WATERLOO.

For their distinguished conduct on the 18th of June, 1815, authority was now granted to the Royal Regiment of Dragoons to bear on their guidons and appointments the word "Waterloo." Every officer and soldier present in the battle received a silver medal, and to the subaltern officers and soldiers was allowed the privilege of reckoning two years' service for that day towards increase of pay and pension. It was not, however, until the 2nd of May, 1838, that the regiment at length received permission to wear an "Eagle," in commemoration of that taken at Waterloo twenty-three years before by Captain Kennedy Clark, the particulars of which interesting episode require to be detailed at greater length; and it will be seen that it was not delay only, but difficulty, which that brave officer experienced in establishing his claim.

The following letter to General the Marquis of

Anglesey, K.G., with the memorandum it inclosed, contained Captain Clark's statement of this affair :—

"NORWICH, 10*th June*, 1817.

"MY LORD,

"I am induced by the interest you have invariably taken in those officers who have had the advantage of serving under your Lordship's command, though personally unknown, to address you under the following circumstances. I had the honour to serve under your Lordship's orders at the battle of Waterloo, where I commanded the centre squadron of the Royal Dragoons in the brigade of the late Sir W. Ponsonby. While performing this duty, I perceived, and led my squadron against, an enemy's "Eagle," the bearer of which (an officer) I ran through the body, in consequence of which the "Eagle" of the 105th Regiment was taken, it falling across me against the horse of Corporal Stiles. I ordered him to carry it at once to the rear, which he did, while I remained in command of my squadron. H.R.H. the Commander-in-Chief has been pleased to reward the corporal with an ensigncy, but has hitherto declined bestowing any mark of his approbation on me, in consequence of my not having been fortunate enough to procure the recommendation of his Grace the Duke of Wellington.

"I inclose a copy of my own statement made at Brussels while confined by my wounds, together with the evidence of two eye-witnesses, taken by Lieutenant-Colonel Radclyffe. Their evidence has not yet been submitted to his Grace, and I feel hopes that, if sanctioned by your Lordship's recommendation, he may be pleased to receive it favourably.

"I have had the honour of being twice under your Lordship's command; have served almost fifteen years in the British cavalry, nearly five of which I was engaged in active service in the Peninsula. I had two horses killed under me, and received two wounds at Waterloo, the second of which obliged me to leave the field about seven o'clock in the evening. Your Lordship's severe wound and return to England prevented me making any application to you at the time, nor should I now trouble your Lordship with the above details (for the tediousness of which I beg to apologise) did I know any other mode of having the transaction cleared up, having been since informed that an officer (Lieutenant Bridges), who was absent during the action, placed the name of Corporal Stiles on the 'Eagle' at Brussels, making no mention of me.

"I have the honour to be, my Lord,

"Your Lordship's most obedient humble servant,

(Signed) "A. K. CLARK,

Captain, Royal Dragoons.

"To the most Honourable the MARQUIS OF ANGLESEY, G.C.B."

"*Memorandum relating to the Capture of the 'Eagle' of the* 105*th Regiment of French Infantry at the Battle of Waterloo, Sunday,* 18*th June,* 1815.

"When my squadron (the centre one) of the Royal Dragoons had advanced 200 or 300 yards beyond the second ledge, and the first line of French infantry had been broke, I perceived, a little to my left, an enemy's 'Eagle' amongst the infantry, with which the bearer was making every exertion to get off towards the rear of the column. I immediately rode to the place calling out to 'Secure the colour!' and at the same time, my horse reaching it, I ran my sword into the officer's right side, who carried the 'Eagle,' who staggered and fell forwards, but I do not think he reached the ground, on account of the pressure of his companions. I immediately called out a second time 'Secure the colour; it belongs to me.' This was addressed to some men who were behind me at the time the officer was in the act of falling. As he fell with the 'Eagle' a little to the left, I was not able to catch the standard so as to hold it. Corporal Stiles and some other men rushed up to my assistance, and the standard was in an instant in the corporal's possession, it falling across him as he came up on my left, before it reached the ground.

(Signed) "A. K. CLARK,
Captain, Royal Dragoons."

Notwithstanding these representations, which with the corroborative evidence were submitted to H.R.H. the Commander-in-Chief and to the Duke of Wellington, they were not favourably entertained, and it was not until the 30th of April, 1839, that as a public and distinguished recognition of his conduct at the battle of Waterloo, Captain, now Lieutenant-Colonel Clark Kennedy, C.B., commanding the 7th or "Princess Royal's" Regiment of Dragoon Guards, was authorised, in addition to the name and arms of Kennedy, to bear the following honourable augmentation in remembrance of this remarkable action, viz. :—

"Upon a chief ermine the representation of a French 'Eagle' and flag, with the inscription 'L'Empereur

Napoleon au 105^{e} Regiment, thereon, and a sword disposed salterwise, and over the same the word 'Waterloo,' being commemorative of the capture of the Eagle at Waterloo, and the sword used on that occasion by the petitioner; and upon wreaths of his liveries are set crests, on the dexter a soldier of the 1st or Royal Dragoons, holding in his right hand a sword proper and in his left a French 'Eagle' with a tricoloured flag having thereon the number 105, and on the sinister a dolphin azure, and upon an escroll underneath this motto— '*Avise la Fin.*'

"30*th April*, 1839."

Lieutenant-General Sir A. Kennedy Clark died in 1844, a K.C.B. and Colonel of the Royal Scots Greys.

Francis Stiles, the corporal of the Royal Dragoons who seems to have been brought forward very needlessly and indeed unjustly, was appointed to an Ensigncy in the 6th West India Regiment on the 23rd of April, 1816, was placed on half-pay on the 20th of December, 1817, and died on the 9th of January, 1828.

The following were the officers who received medals for Waterloo, in addition to which Colonel Arthur B. Clifton received the Companionship of the Order of the Bath, the Second Class of the Russian Order of St. Anne, and the Fourth Class of the Order of Wilhelm of Holland. Major and Brevet Lieutenant-Colonel Philip Dorville, the Companionship of the Order of the Bath:—

MEDALS.

Captains.—Brevet-Major Charles E. Radclyffe, Alexander Kennedy Clark, Paul Phipps.

Lieutenants.—Henry Robert Carden, Sigismund Trafford, George Gunning, Townsend Richard Keily, Samuel Wyndowe, Cornthwaite Ommaney, Charles Blois, Stephen Goodenough.

Cornets.—C. B. Stephenson, Honourable John Massey, Quartermaster W. Waddell, Surgeon George Steed, Veterinary Surgeon W. Ryding.

We left the Royal Regiment of Dragoons on the 25th of January, 1816, in the barracks at Ipswich, whence, towards the end of August, 1817, it marched for Scotland, to be there stationed at Glasgow, Hamilton, Ayr, Dumfries, and Stirling. In June, 1818, the regiment embarked at Port Patrick for Ireland, and, landing at Donaghadee, proceeded to Longford, Ballinrobe, Sligo, Roscommon, and Dunmore. In November a reduction of 8 sergeants, 96 men, and 56 horses was made in the establishment.

In June, 1819, proceeding to Dublin, the regiment remained there until, in August of the following year, it embarked for England, and, landing at Liverpool, moved to Manchester, Oldham, Ashton, and Altrincham.

On the 20th of February this year, 1820, a submission to the King was drafted for the suppression of the bayonet in the heavy cavalry, a question which had been raised in the year preceding, and in consequence it was discontinued.

On the 19th of March, 1821, the Royals commenced their march to Radipole barracks near Weymouth, whence a number of parties were detached on revenue service and for the seizure of smuggled goods. While on this duty the regiment received upwards of £200. In September, the establishment was further reduced to 6 troops of 3 officers, 3 sergeants, 1 trumpeter, 1 farrier, 50 rank and file, and 42 horses each.

The regiment marched on the 13th of June, 1822, from the west and south-western districts to Richmond and other places near London ; and on the 6th of July it was reviewed on Wormwood Scrubs by Field-Marshal His Royal Highness the Duke of York, K.G., two days after which it moved to Canterbury, detaching troops and parties on revenue duty.

Having called in their detachments, the Royal Dragoons left Canterbury on the 1st of July, 1823, for the cavalry barracks, Regent's Park, London, where, on arrival, they took the "King's" duty, in the absence of the Household Cavalry, moved to quarters near Hounslow, preparatory to a review on the 15th of the month, when the Royal Regiment of Dragoons furnished a guard of honour to Field-Marshal His Royal Highness the Commander-in-Chief, as well as a squadron to keep the ground. On the following day, being relieved on the King's duty, they marched for York, arriving there on the 29th of the month.

From York the Royals marched on the 24th of May, 1824, for Scotland, and there occupied Piershill Barracks, Edinburgh, with detachments to Perth, Cupar Angus, and Forfar. At the calamitous fire which broke out in the Parliament Square in Edinburgh, in the month of

November, the regiment was employed for three successive nights preserving order, protecting property, and rendering assistance to the unfortunate sufferers; and the dismounted men, with the barrack engine, aided materially in extinguishing the flames in the Tron Church. Their services on this occasion were commended in a general order by Major-General Sir Thomas Bradford, commanding the forces in North Britain, and in a vote of thanks from the Lord Provost, the magistrates, and the Town Council of Edinburgh.

In the early part of March, 1825, the regiment moved to Glasgow and Hamilton, whence, in the following month, it embarked for Ireland, where again landing at Donaghadee, it marched to Dundalk and Belturbet, whence several strong parties were required for the conduct of specie, the coinage of the two kingdoms having been at this time assimilated.

On the 30th of March, 1826, the Royals moved to Dublin.

On the 24th of February, 1827, died in Connaught Square, London, an officer whose services in the Royal Dragoons had established for him a well-deserved reputation—Lieutenant-Colonel Radclyffe, Brigade-Major to the cavalry in Great Britain, at the age of 53. This remarkable officer had served in most of the campaigns of the late revolutionary war, commencing under H.R.H. the Duke of York in Flanders in 1793, and ending with the great victory of Waterloo in 1815, where he received a severe wound from a musket ball lodging in the knee, and which, from the impossibility of extraction, continued to be a cause of great suffering to the end of his life. His lieutenant-colonelcy dates from that glorious day, previous to which he had been present in the Peninsula and south of France, as Brigade-Major of cavalry, at the battles of Busaco, Fuentes d'Onor, Salamanca, Vittoria, the blockade of Pampeluna, investment of Bayonne, besides various engagements of less note. Up to the battle of Toulouse on the 10th of April, 1814, Lieutenant-Colonel Radclyffe had held the appointment of Brigade-Major, but from that date he was promoted to the higher one of Assistant-Adjutant-General to the cavalry, in which capacity he accompanied its march through France to England. So entirely was his mind engrossed by his profession, that almost his last words, but two hours before his death, to the inquiries of his physician were—"I am retreating, retreating, retreating; I cannot advance." He was a dexterous swordsman, an accomplished officer, and an able tactician, on which subjects he had published a small work. A warm and sincere friend, a conscientious Christian, and brave man, Lieutenant-Colonel Radclyffe lived universally and highly respected, as he died sincerely lamented.

In April, 1827, the Royals moved to Newbridge, and thence in October following to Cork, Fermoy, and Bandon. In March, 1828, the whole assembling at Cork marched into barracks at Ballincollig.

On the 26th of April, 1829, the regiment commenced

its march to Dublin, where early in May it embarked for England, and landing at Liverpool, went into billets in the town of Manchester, the barracks there having been pulled down for the purpose of reconstruction. On the 10th of June, Colonel Clifton, whose services in command of the regiment had been so long and so conspicuous, exchanged to half-pay with Lieutenant-Colonel Charles Somerset, unattached.

During their stay in Manchester, the Royals were frequently called upon for pickets and parties for the prevention of riots and violation of the law by the operatives, who were generally in a state of serious disaffection. Detachments were also sent to Blackburn and Bolton with the same object.

The death of General Garth, on the 16th of November this year, placed the colonelcy of the Royal Regiment of Dragoons at the disposal of His Majesty King George IV., who was pleased to confer it upon Lieutenant-General Lord R. Edward Somerset, G.C.B., from the 17th Lancers, upon which occasion, in a letter dated "Royal Lodge, November 22nd, 1829," and addressed to the Viscount Hill, commanding the army in chief, referring to certain arrangements submitted by his Lordship, His Majesty observes :—

"MY DEAR LORD,

"Subsequent to the receipt of your letter of this day, your letter of the 19th reached my hand, and with reference to the recommendation therein preferred,

I have to observe to you that I consider it essential for the service that the Royal Dragoons should ever be held by an officer of rank, notwithstanding the two successive precedents to the contrary; and as Lieutenant-General Vandeleur has already a regiment of equal emolument, I prefer that Lieutenant-General Lord Edward Somerset should be removed from the 17th Lancers to the Royals."

Early in June, 1830, the regiment moved to Norwich and Ipswich, the establishment at the same time being reduced to 270 rank and file.

On the 26th of June died King George IV., and one of the first acts of his successor, William IV., was to order the abolition of the moustaches in the Royal Horse Artillery, and throughout the cavalry, with the exception of the Household and the Hussars. During this autumn the agricultural labourers excited by mischievous and designing individuals, and also, as was supposed, irritated by the increased introduction of steam machinery in several counties, committed numerous acts of incendiarism and destruction of property; and in consequence the regiment was called upon for a number of detachments, to assist the civil power in suppressing these outrages, and a resolution of thanks from the magistrates of Norfolk, acknowledging the very effective service rendered by the officers, non-commissioned officers, and privates of the regiment, was forwarded by the Lord-Lieutenant of the county to General Viscount Hill, who was pleased to express his satisfaction on being presented with so honourable a testimonial of their conduct.

The regiment continued in these quarters the whole of the year 1831, and on the 20th of March, 1832, the head-quarters left Norwich for Canterbury, arriving there on the 24th of the month. On the 13th of May a route was received for the immediate march of 4 troops to Dartford, where, on arrival, orders were received to proceed to Woolwich, there to remain in case of their services being required in London. On the 25th of the month they returned to Canterbury.

On the 27th of July, the regiment was reviewed by the Right Honourable the General Commanding-in-Chief, who expressed himself well pleased with their appearance.

On the 6th of August, a squadron, consisting of 2 captains, 2 subalterns, 4 sergeants, and 100 rank and file, with the Royal Standard, under the command of Major Martin, marched to Colnbrook, in the neighbourhood of Windsor, in order to furnish, by express command of His Majesty, the guard of honour upon the occasion of a review, to be held at Windsor on the 13th instant, at which it was the King's intention to present an additional standard to the Royal Regiment of Horse Guards. On the following day, 4 troops under the command of Lieut.-Colonel Somerset left Canterbury for Hyde Park Barracks, London, for the purpose of doing duty at the Horse Guards during the absence of the Life Guards, moved to Windsor, and while in London, the office of Silver Stick was held by Lieutenant-Colonel

Somerset. At the review on the 13th instant, Major Martin, commanding the guard of honour, received from His Majesty the decoration of the Royal Hanoverian Guelphic Order. Upon being relieved by the Life Guards on the 14th instant, the Royals returned to Canterbury.

In consequence of the general elections for the county and city, the regiment on the 7th of December was detached as follows: head-quarters and 3 troops to Margate; 2 troops to Faversham; and 1 troop to Ospringe; the whole returning to Canterbury on the 25th of the month.

The head-quarters of the Royal Dragoons left Canterbury on the 4th of March, 1833, for Dorchester, whence a troop was subsequently detached to Weymouth, and another to Christchurch and Winchester, and early in the following year the regiment moved to Brighton, with a squadron at Canterbury.

On the 4th of July, 1834, the Adjutant, Mr. Kelly, having gone out riding on the Downs, was discovered lying on the ground insensible, in which state he died, and was buried in Preston churchyard.

On the 1st of January, 1835, having received a sudden order for Ireland, in consequence of a general election expected there, and being relieved by the Life Guards, the Royal Dragoons left Brighton for embarkation at Bristol, and in the course of this march, owing to the severity of the weather and the condition of the roads, some difficulty and several accidents occurred both in England and after arrival in Ireland, where the regiment landed at Dublin on the 18th of January, the head-quarters proceeding direct to Newbridge, with detached troops at Kells, Navan, Carlow, and Kilkenny.

On the 28th of May, died very suddenly at Ripley, in Surrey, Lieutenant-Colonel Somerset, who the day following was succeeded by Major Martin, K.H., who had been several years in the 2nd Life Guards, with which corps he had served in the Peninsula, the south of France, and at the battle of Waterloo.

On the 30th of the same month, Lieutenant the Honourable Joshua Vanneek died in Newbridge Barracks from the effects of an accident which had happened to him a few days previous.

On the 31st of March, 1836, the colonelcy of the Royal Regiment of Dragoons having become vacant by the transfer of Lord R. E. Somerset to the 4th Queen's Own Light Dragoons, it was conferred upon Major-General the Honourable Sir Frederic Cavendish Ponsonby, K.C.B., K.C.H., G.C.M.G., from the 86th Foot.

The regiment left Newbridge for the Royal Barracks, Dublin, on the 7th of May, and in the spring of this year the so-called "Roman" helmet was replaced by one of brass, for the officers gilt, with a movable bearskin crest. Several minor changes of dress also took place, and the blue uniform of the band, which, as the Royal Regiment of Dragoons, they had worn for many years, was discontinued by order of King William IV.

Major-General the Honourable Sir Frederic Ponsonby dying on the 10th of January, 1837; he was succeeded in the colonelcy of the Royals by Lieutenant-General the Right Honourable Sir Richard Hussey Vivian, Bart., K.C.B., G.C.H., from the 12th Royal Lancers, on the 20th of the month.

The accession to the throne of Her Majesty Queen Victoria on the 20th of June this year, occasioned much employment for the troops in Ireland, and particularly for the Royals, during the months of July and August, in assisting the civil power in the course of the elections, but on the 20th of September the regiment finally quitted Dublin for Cork and Ballincollig.

Major-General Sir Amos Norcott, K.C.H., C.B., commanding the Cork District, died, and was buried with military honours, at which the regiment assisted on the 8th of January, 1838, Major Banner, of the 92nd Gordon Highlanders being similarly interred on the same day.

A Horse Guards order of the 2nd of May, 1838, communicated to the regiment Her Majesty's gracious permission for the Royal Dragoons to assume on their guidons and appointments the "Eagle," in commemoration of their having captured that of the 105th regiment of the French Line at the battle of Waterloo.

On the 5th of the month came a sudden order for the embarkation of the regiment for Liverpool, where, on arrival, the head-quarters and 1 troop proceeded to Sheffield, the remaining 5 troops being dispersed throughout the Northern District, commanded by Major-

General Sir Charles Napier, and at this time much disturbed by meetings of the Chartists, headed by Feargus O'Connor. For some weeks a small force under Lieutenant-Colonel Martin, K.H., consisting of a troop of the Royal Horse Artillery, 1 troop of the Royal Dragoons, and some companies of the 20th regiment of infantry, was stationed at Ashton-under-Lyne, which neighbourhood was greatly disaffected. The regiment, concentrating at Sheffield, was much harassed during the winter, so much bad feeling prevailing in the town; where, among other modes of annoyance, the people threw down in the streets what are called "crows' feet," to injure the horses. While here the horse artillery and cavalry recommenced wearing the moustaches.

At a meeting held on the 5th of February, 1840, the services of the regiment were publicly acknowledged in a vote of thanks by the magistrates and the commissioners for the improvement of the town of Sheffield, who upon the same occasion presented Lieutenant-Colonel Martin with a handsome piece of plate.

By order from the Horse Guards of the 29th of February, 1840, the pistols were abolished in the heavy cavalry, excepting for troop-sergeant-majors and trumpeters, those in the light cavalry, with the exception of the Lancers, having been equally suppressed in the year preceding.

On the 7th of May the regiment commenced its march for Glasgow and Hamilton, and on the 9th of June it was inspected by Major-General Lord Greenock, K.C.B., commanding in North Britain, who expressed himself highly satisfied with its appearance.

On the 19th of September, it was inspected by Major-General Sleigh, C.B., Inspecting General of Cavalry, who expressed himself equally pleased.

One squadron marched on the 5th of April, 1841, for Newcastle-on-Tyne; 1 troop on the 12th for Carlisle; and the head-quarters on the 26th of May for Leeds, detaching 1 troop to Bradford.

On the 13th of September, Major-General Sleigh, C.B., inspected the regiment, with which he was well pleased.

The troops from Carlisle and Newcastle-on-Tyne arrived on the 12th of April, 1842, at Manchester, and the head-quarters and the troop from Bradford on the 12th of May, in which month a troop was detached to Congleton, rejoining head-quarters in August, when another troop went to Burslem in aid of the civil power, but returned in a few days.

On the 9th of August serious disturbances broke out in the town and neighbourhood of Manchester, a general turn-out of the operatives taking place, which continued for some weeks, during which the military were constantly and actively employed. On the 18th of the month, Lieutenant-Colonel Martin was detached to Ashton-under-Lyne, in command of a troop of horse artillery, a squadron of the Royals, and some detachments of infantry, to the assistance of the magistrates, where he remained several weeks.

General Lord Vivian, dying on the 20th of August, was succeeded in the colonelcy of the Royal Regiment of Dragoons by General Sir Arthur B. Clifton, G.C.B., K.C.H., from the 17th Lancers.

With reference to the conduct of the military throughout these troubles in the North-Western District, the following extracts and correspondence refer particularly to the Royal Dragoons.

"GENERAL ORDER.

"HORSE GUARDS, *8th Oct.*, 1842.

"The Commander-in-Chief has great satisfaction in publishing the expression, conveyed to him by the Secretary of State, of Her Majesty's most gracious approbation of the services performed as well by the Yeomanry as Her Majesty's troops."

"WHITEHALL, *23rd Sept.*, 1842.

"MY LORD DUKE,

"I have the honour to inform your Grace that I have received the Queen's commands to express Her Majesty's high approval of the good conduct, exemplary forbearance, and steadiness, of the military employed in support of the civil authorities, during the disturbances which have unhappily prevailed in many of the northern and midland counties, and to communicate to Lieutenant-General Sir Thomas Arbuthnot, Commanding-in-Chief in these counties, to Major-General Sir William Warre, Major-General Brotherton, and Colonel Thorn, to the officers, non-commissioned officers, and privates, the expression of Her Majesty's approval.

"I have, &c.,

(Signed) "R. T. GRAHAM.

"His Grace the COMMANDER-IN-CHIEF."

The regiment was inspected on the 14th of October by the Inspecting General of Cavalry, Major-General the Honourable E. H. Lygon, C.B., and Lieutenant-Colonel Martin having retaken the command, received the following letter from the magistrates of Ashton-under-Lyne, and Staleybridge.

"ASHTON-UNDER-LYNE, *November*, 1842.

"SIR,

"We, the undersigned magistrates acting in the district and neighbourhood of Ashton-under-Lyne, feel ourselves called upon by duty no less than inclination to express to you our high sense of the important services rendered by you in aiding us in the preservation of the public peace during the late tumultuous scenes of riot and disturbance; and we are most anxious to express to you our thanks for the energy and promptitude evinced by you in attending to our suggestions on all occasions.

"We also beg you, Sir, to be good enough to convey to the officers and privates of the Royal Dragoons, the Royal Horse Artillery, and the several detachments of infantry under your command, our thanks for their important services; and we had the greatest pleasure in noticing the orderly and correct deportment in quarters of the non-commissioned officers and privates of these distinguished corps, which has left on the minds of this community a lasting impression of regard and respect.

"We have the honour to be, SIR,

"Your very obedient Servants,

(Signed) "JOSHUA T. R. EVANS,
"JAMES JOWITT,
"JAMES LORD,
"JONAH HARKOFS,
"JOHN GRIMSHAW,
"RALPH CUSEY,
"WILLIAM SIDEBOTTOM,
"JOHN SIDEBOTTOM,
"ROBERT DE HOLLINGSWORTH.

"LIEUTENANT-COLONEL MARTIN, K.H.,
Commanding Royal Dragoons."

"HEAD-QUARTERS, N.W. DISTRICT,
CHESTER, 18*th Dec.*, 1842.

"SIR,

"Major-General Sir William Warre has great satisfaction in communicating to you, and to request you will promulgate to the regiment under your command, the following extract from a letter he has received from Lieutenant-General Sir Thomas Arbuthnot, in which he is desired to inform you that the Lieutenant-General Commanding did not fail to bring to the Commander-in-Chief's notice the very efficient state in which he found the Royal Dragoons, and the very soldierlike and cheerful manner in which they performed their duties when called out in aid of the civil power during the late disturbances. And that the Duke of Wellington has derived much satisfaction from the perusal of these reports generally, as well as from the assurances contained in the Lieutenant-General's letter, of the good order of the corps."

(Signed) "G. C. MUNDY, *Major*,
Brigade-Major.

"The Officer Commanding Royal Dragoons, Manchester."

The head-quarters of the Royal Dragoons left Manchester for Birmingham on the 24th of April, 1843, detaching 1 troop to Coventry, 1 to Dudley, and 1 to Newcastle-under-Lyne. The regiment was inspected by Major-General the Honourable E. P. Lygon, C.B., on the 29th of August, and leaving Birmingham on the 17th of October, embarked at Liverpool on the 25th for Dublin, whence, on the 26th, it marched to Newbridge.

On the 24th of April, 1844, the regiment moved to Dublin, where, on the 30th of May, it was inspected by Major-General Wyndham, and again by him on the 25th of September following.

Major-General Wyndham inspected the Royals on the 12th of May, 1845, and on the 16th of September the head-quarters and 1 troop marched to Dundalk, detaching a squadron to Belturbet, 1 troop to Belfast, 1 to Monaghan, and 1 to Inniskillen.

On the 2nd of October, Major-General Sir George Berkeley, K.C.B., commanding the Belfast District, made an inspection of the regiment, which, on the 13th of May, 1846, moved to Cork; and while on the march, in consequence of the disturbed condition of a portion of the Limerick and Tipperary districts, 1 troop was halted at Clonmel, and another at Carrick-on-Suir, and were employed in convoying flour from the interior of these districts for exportation to the English markets. These disturbances, which at first threatened to be serious, were suppressed by the capture of most of the ringleaders, when the country became quiet and settled.

On the arrival of the regiment at Cork on the 27th of May, it was stationed as follows: head-quarters and 2 troops at Cork; the recruits and young horses at Ballincollig; 1 troop at Bandon; and a troop each at Clogheen, Waterford, and Clonmel.

On the 1st of June, Major-General Turner, commanding the Cork District, inspected the regiment, and again on the 5th of November. It was inspected by Major-General Turner on the 27th of May, 1847, and on the 16th of September the regiment marched to Cahir, arriving on the 16th, and detaching troops to Limerick, Rathkeale, and Newcastle. On the 22nd of October it was inspected by Major-General Napier, C.B.

On the 1st of April, 1848, in consequence of Horse Guards orders dated respectively the 30th of July and 3rd of December, 1847, a new pattern brass helmet, for the officers gilt, and a new pattern coatee with shorter skirts, were taken into wear.

On the 17th of May, the regiment was inspected by Major-General Napier, C.B., and leaving Cahir on the 5th of June for Newbridge, it embarked on the 13th at

N

Dublin for Liverpool, where on landing the head-quarters and 1 troop proceeded to York, 1 troop to Newcastle-on-Tyne, a squadron to Leeds, 1 troop to Halifax, and 1 to Bradford.

The regiment was inspected on the 18th of November by Major-General Thorne, C.B., commanding the Northern District.

On the 3rd of September, 1849, the Royal Regiment of Dragoons was inspected by Major-General Brotherton, C.B., Inspecting-General of Cavalry, and on the 22nd of October by Major-General Thorne, C.B.

On the 20th of April, 1850, the head-quarters and 2 troops left York for Nottingham, occupying the barracks there, and detaching 3 troops to Sheffield, and 1 to Loughborough. The regiment was inspected on the 3rd of June by Major-General Sir William Warre, commanding the District, and on the 3rd of August by Major-General Brotherton, C.B.

Major-General Sir William Warre inspected the regiment on the 27th of March, 1851, and on the 1st of May it left Nottingham for Barnet, near London, arriving there on the 9th, to remain pending the result of the first Great Exhibition opened in London on the 1st of May.

The head-quarters and 2 troops were billeted in Barnet, 2 at Enfield, 1 at Potter's Bar, and 1 at Whetstone. The dismounted men and heavy baggage had been sent to the barracks in Northampton on the 10th instant.

The 6 troops of the Royal Dragoons assembling in review order at Barnet, were there inspected by Major-General H.R.H. the Duke of Cambridge, K.G., on which occasion His Royal Highness was pleased to pass a high encomium on their appearance.

Having been inspected by Major-General Brotherton, C.B., on the 24th of July, the regiment on the 5th of August marched for Brighton, arriving on the 9th, and detaching 2½ troops to Christchurch.

On the 14th of April, 1852, the Royal Dragoons paraded in review order for the inspection of their Colonel, Lieutenant-General Sir Arthur B. Clifton, K.C.B., K.C.H., and with new guidons, the old ones now replaced having been in wear about 24 years. The motto of the regiment, *Spectemur Agendo*, did not appear upon the new guidons by order of the Inspector of Colours, but it has since been restored.

The regiment left Brighton for Dorchester on the 22nd of June, and when *en route* at Ringwood, received orders to halt, and on the 29th it was ordered to Birmingham, where it arrived on the 8th of July, detaching thence 4 troops to Coventry under Major J. Yorke.

On the 30th of the month, leaving Birmingham for Manchester, the Royals arrived there on the 8th of August, detaching a troop at Preston.

On the 12th, the regiment was inspected by Major-General H.R.H. the Duke of Cambridge, K.G., Inspecting-General of Cavalry.

Colonel Marten, K.H., having commanded the Royal Regiment of Dragoons for the long period of 18 years, retired upon half-pay on the 4th of February, 1853, being succeeded by Major John Yorke. This much respected officer died on the 23rd of November, 1868, a Lieutenant-General and Colonel of the 6th or Inniskillen Dragoons.

On the 1st of September, Major-General H.R.H. the Duke of Cambridge, K.G., made an inspection of the regiment.

CHAPTER VII.

THE CRIMEA.

IN view of the impending war with Russia to be proclaimed by England and France in support of Turkey, and in the interest of the balance of power in Europe, the Royal Regiment of Dragoons was ordered to hold itself in readiness for immediate service in the East, the scene of expected operations being the Crimea, and in consequence, on the 20th of March, 1854, the regiment concentrated at Manchester, for the purpose of organising the service and depot troops, which was effected without calling for either man or horse from any other corps.

War being declared on the 27th of March, on the 9th of May the head-quarters left Manchester, and the day following embarked at Liverpool in two transports, the *Gertrude* and the *Peruvian*. On board the former were Lieutenant-Colonel Yorke, Lieutenant Coney, Cornet Robertson, Surgeon Barron, 3 sergeants, 1 trumpeter, 1 farrier, 50 rank and file, 2 women, 9 officers' horses, 45 troop horses; in the *Peruvian*, Captain Elmsall, Cornet

Glyn, Cornet Sandeman, Assistant-Surgeon Gorringe, 4 sergeants, 1 trumpeter, 1 farrier, 43 rank and file, 2 women, 8 officers' horses, 39 troop horses. After a tedious passage this portion of the regiment landed at Varna on the 28th of June, and encamped at the head of Varna Bay.

A second detachment embarked at Liverpool on the 20th of May in the transports *Arabia* and *Rip van Winkle,* under the command of Major Wardlaw, and was composed as follows: Captain Campbell, Lieutenant Charlton, Adjutant Webster, Cornet Wartopp, Veterinary-Surgeon Post, 7 sergeants, 1 trumpeter, 2 farriers, 116 rank and file, 4 women, 16 officers' horses, 113 troop horses. The *Arabia* reached Varna on the 7th, and the *Rip van Winkle* on the 14th of July, and joined the head-quarters in camp.

On the 28th of the same month a third detachment embarked at Liverpool under Captain Storks in the transport *Coronetta,* comprising Lieutenant Pepys, Quartermaster Scott, 1 sergeant, 1 trumpeter, 1 farrier, 48 rank and file, 7 officers' horses, 41 troop horses, and landed at Varna on the 10th of July, where it was found that glanders had broken out among the horses, caused by the imperfect ventilation of the vessel. These horses were immediately encamped by themselves on the heights above the Bay of Varna, and it was subsequently found necessary to destroy 25 of them.

The fourth and last detachment, embarking at Liverpool on the 30th of May on board the transport *Conrad,* under the command of Captain Sykes, consisted of Lieutenant

Basset, 2 sergeants, 1 farrier, 15 rank and file, 5 officers' horses, 10 troop horses, and landed at Varna on the 14th of July.

By authority dated "War Office, 13th of June, 1854," the regiment was increased to the following strength from the 10th of May previous :—

	Sergeants.	Farriers.	Corporals.	Privates.	Horses.
4 Service Troops	42	4	12	261	250
2 Depot Troops	12	3	11	144	71
Establishment . . .	54	7	23	405	321

On the 30th of June a troop under the command of Captain Elmsall received orders to march to Devna, and thence to detach a party of 1 sergeant and 8 privates, under Cornet Glyn, for the purpose of escorting forage to the Light Cavalry Brigade, which, commanded by Brigadier-General the Earl of Cardigan, had pushed on towards the Danube, in order to reconnoitre the position of the enemy. This escort overtook the Light Brigade about 30 miles from the Danube, and was then directed to proceed to Shumla, whence it subsequently rejoined the troop at Devna on the 11th of July.

On the 10th of July a detachment under Major Wardlaw left Varna for Devna, being followed on the 17th by the head-quarters of the regiment, which then was brigaded with the 5th Dragoon Guards, under Brigadier-General the Honourable J. Y. Scarlett; but cholera having appeared in that regiment, as well as in the Light Infantry Division encamped on the other side of the plain, it was considered advisable to break up the general encampment, and on the 29th of July the Royal Dragoons moved to Karra. On the 3rd of August, however, occurred the first fatal case of cholera, followed on the 8th and 9th by the death of 5 men and 1 woman, when the camp being immediately changed to the proximity of the village, the disease disappeared.

A fire having occurred at Varna, which destroyed the chief part of the commissariat stores of barley, the Royal Dragoons received no forage of any description from the commissariat for several days, and with difficulty procured a few loads of grass and sheaf barley from the villagers.

Leaving Karra Pass on the 16th of August, the Royals rejoined the 5th Dragoon Guards, and, under the command of Brigadier-General the Honourable J. Y. Scarlett, encamped in the evening about three miles from Devna, the horses having been without food the whole day. A foraging party returned at night with a few loads of sheaf barley. Resuming the march on the following day, the regiment encamped at night near the lake of Varna, the horses receiving a sheaf of barley each. During the night a dragoon died of cholera.

On the following morning the regiment proceeded to the heights near the Adrianopolis road, and there joined the Brigade, which now consisted of the 4th and 5th Dragoon Guards, with the Royal and Inniskillen Dragoons. In the course of the month a remount of 36 horses joined the regiment.

On the 7th of September, the combined British

and French expeditions sailed from the Bays of Varna and Baltchik for the Crimea, the Light Cavalry only accompanying the British force. The Heavy Cavalry followed on the 24th, the head-quarters of the Royal Dragoons under Lieutenant-Colonel Yorke sailing on the 25th in the transport *Rip van Winkle*, with Captain Campbell, Cornet Hartopp, 5 sergeants, 1 trumpeter, 1 farrier, 58 rank and file, 1 woman, 8 officers' chargers, and 61 troop horses.

Major Wardlaw, Captain Storks, Lieutenants Charlton and Pepys, Cornets Robertson and Sandeman, Veterinary-Surgeon Post, 10 sergeants, 2 trumpeters, 1 farrier, 88 rank and file, 1 woman, 18 officers' chargers, and 92 troop horses, embarked in the *Wilson Kennedy*.

Captain Elmsall, with Captain Sykes, Lieutenants Coney and Barrett, Cornet Glyn, Assistant-Surgeon Gorringe, Quartermaster Scott, 8 sergeants, 1 trumpeter, 1 farrier, 81 rank and file, 5 women, 15 officers' chargers, and 82 troop horses, embarked in the *Pride of the Ocean*. The transports, towed by steamers, sailed under convoy of H.M. steamer *Spiteful*; but they had scarcely left the harbour when a tremendous storm came on, the hawsers parted, the *Wilson Kennedy* was driven down to the Bosphorus, losing 99 officers' and troop horses. The *Rip van Winkle* lost 43, and the *Pride of the Ocean* 8. The *Wilson Kennedy* landed 11 horses saved at Scutari, where the officers and men were trans-shipped to the *Cambria*, and landed at Balaclava, in the Crimea, on the 4th of October. The head-quarters disembarked on the day following, when the regiment encamped with the cavalry division, commanded by Lieutenant-General the Earl of Lucan, on the plain of Balaclava, where the Heavy Brigade was increased by the Scots Greys.

Marshal St. Arnaud, Commander-in-Chief of the French army, dying of cholera on the 29th of September, 1854, was succeeded by Marshal Canrobert.

On the 18th of October the regiment received 75 horses from the Light Brigade.

The cavalry division, turning out every morning before daylight, was kept continually on the alert by the frequent demonstrations of the Russians in the direction of the Tchernaya Valley and the village of that name. On the 19th the cavalry encampment was shifted higher up the plain, the right resting on the road to Sevastopol, and here also the troops continued to be harassed by the enemy, being in constant readiness to turn out, and upon one occasion they remained the whole night under arms.

On the 25th a division of the Russian army, commanded by General Leprandi, advancing from the village of Tchergoum, attacked the allied position in front of Balaclava. The low range of heights which crosses the plain, at the extremity of which lies the town, was protected by four redoubts, defended by Turkish troops, the principal of which, adjoining the Kamara village, was the first attacked and carried before 8 a.m., and little resistance being offered by the other three, these also were soon taken possession of, the Turks retreating in disorder. From the commencement of the attack, and during their retreat, the Turks were supported by Captain Maud's troop of Royal Horse Artillery, and by the cavalry division, who, in doing so, were exposed to the fire of heavy guns and riflemen, the Royals losing 1 man and 2 horses killed.

The redoubts having been thus captured, the Russian cavalry, consisting of 16 squadrons of Hussars, 10th and 12th, the Duke of Saxe Weimar's regiment, the Ural Cossack regiment, and 3 sotnias of the 53rd Don Cossacks, supported by the 12th Battery of Horse Artillery and the 3rd Don Cossack Battery, now moved forward on the heights, and attacked the Heavy Cavalry Brigade of Major-General the Honourable J. Y. Scarlett, formed in two lines, the first consisting of the Royal Scots Greys and Inniskillen Dragoons, the second of the 4th and 5th Dragoon Guards, the Royal Dragoons being in echelon in rear to the left of the Greys. The first line charged, but the enemy, in greatly superior numbers, outflanking the British, the Royals, led by Lieutenant-Colonel Yorke, inclining to the left and encountering the right of the Russians, checked its advance, threw it into confusion, and finally it went about and galloped off the field. Previous to the charge, 1 man of the regiment was killed by a cannon-shot, 2 men were wounded, and there were 7 horses missing.

After this charge the Heavy Brigade moved forward about 3 miles to the neck of the valley, where the Light Cavalry Brigade had been ordered to attack the

Russian batteries, and in this movement of support, the Royals and Greys, being in the first line, were exposed to a heavy cross fire from guns and infantry in the captured redoubts on both flanks, and here were wounded Lieutenant-Colonel Yorke very severely (his horse also by a grape shot), Captain Elmsall severely, Captain Campbell very severely, and Cornet Hartopp also severely, 2 men and 8 horses wounded; and 1 man wounded died during the night. After the brilliant but disastrous charge upon the batteries by the Light Cavalry led by the Earl of Cardigan, the Heavy Brigade covered their retreat, and remained on the ground till dark.

On the 26th the encampment was moved to a position under the heights before Sevastopol, overlooking the plain of Balaclava, and on the 28th the cavalry again shifted their encampment to the heights before Sevastopol.

With reference to the events of the 28th instant, the following General Order now appeared:—

"HEAD-QUARTERS BEFORE SEVASTOPOL,
"*October 29th*, 1854.

"GENERAL AFTER ORDER, No. 2.

"The Commander of the Forces considers it his duty to notice the brilliant conduct of the Cavalry Division, under the command of Lieut.-General the Earl of Lucan, in the action of the 25th inst. He congratulates Brigadier-General the Honble. J. Y. Scarlett and the officers and men of the Heavy Brigade upon their successful charge and repulse of the Russian Cavalry in far greater force than themselves, and while he condoles with Major-General the Earl of Cardigan and the officers and men of the Light Division on the heavy loss sustained, he feels it to be due to them to place on record the gallantry they displayed and the coolness and perseverance with which they executed one of the most arduous attacks that was ever witnessed, under the heaviest fire and in face of powerful bodies of Artillery, Cavalry, and Infantry.

(Signed) "T. B. B. ESTCOURT,
"*Adjt.-General.*"

From the 1st of October the establishment of the regiment had been augmented as follows: viz. from 399 men to 450, so as to consist of 75 men per troop, and from 321 horses to 372, making 62 per troop.

	Sergeants.	Trmptrs.	Farriers.	Corporals.	Privates.	Horses.
4 Service Troops .	26	4	4	16	299	300
2 Depot Troops . .	8	3	2	7	151	72
Establishment .	34	7	6	23	450	372

On the 5th of November, at the battle of Inkermann, the cavalry were posted in support, the Royal Dragoons being in rear of the French infantry on the heights above the plain of Balaclava, where the Russian army of Leprandi seemed to threaten the town, but as soon as it became evident that this was merely a feint, the regiment was ordered to Inkermann with a troop of Horse Artillery, where it remained until the close of the action.

On the following day the camp of the cavalry was moved a short distance on the heights. On the 15th of the month the establishment of the regiment was augmented as follows: 8 troops, 1 colonel, 1 lieutenant-colonel, 1 major, 8 captains, 8 lieutenants, 8 cornets, 1 paymaster, 1 adjutant, 1 quartermaster, 1 surgeon, 2 assistant-surgeons, 1 veterinary surgeon, 1 regimental sergeant-major, 8 troop sergeant-majors, 1 paymaster-sergeant, 1 armourer-sergeant, 1 farrier-sergeant, 1 schoolmaster, 1 hospital-sergeant, 1 orderly-room clerk, 31 sergeants, 31 corporals, 1 trumpeter-major, 8 trumpeters, 8 farriers, 600 privates, and 520 horses.

	Sergeants.	Trmptrs.	Farriers.	Corporals.	Privates.	Horses.
6 Service Troops .	38	7	6	24	449	450
2 Depot Troops . .	8	2	2	7	151	70
Establishment .	46	9	8	31	600	520

By Horse Guards memorandum dated November 20th, 1854, it was directed that in order to facilitate the recruiting both of men and horses, and to expedite their training, the regiments at home should raise 60 men and 120 horses, to be by them drilled and trained for the 10 regiments serving in the East; the 6th Dragoon Guards being ordered to provide for the Royal Dragoons.

Throughout this month of November, both men and horses suffered severely from constant exposure to the wet and cold weather and insufficiency of forage.

On the 5th of December were published the following General Order :—

"HEAD-QUARTERS BEFORE SEVASTOPOL,
"5*th December*, 1854.

"The Commander of the Forces has the highest satisfaction in communicating to the Army a despatch which he received yesterday from the Minister of War, His Grace the Duke of Newcastle, expressing the Queen's admiration of the conduct of the troops engaged in the action in front of Balaclava, on the 25th of October, 1854.

"WAR DEPARTMENT, 14*th November*, 1854.

"MY LORD,

"I have the honour to acknowledge your Lordship's despatch No. 35, dated before Sevastopol, October 28th, in which you give an account of a battle fought on the 25th of that month in front of Balaclava. I have laid that despatch before the Queen, and I have received Her Majesty's command to express to your Lordship her admiration of the gallantry and conduct of the troops engaged upon that occasion. Her Majesty has learned with deep concern that the repulse of the enemy was not effected without a heavy loss of the Division of Cavalry, more especially of the Light Brigade, but the brilliancy of the charge, and the gallantry and discipline evinced by all, have never been surpassed even by British soldiers under similar circumstances.

"To every officer, non-commissioned officer, and private, engaged in this severe encounter with vastly superior numbers, the Queen desires me to communicate through your Lordship her approval and thanks.

"Her Majesty has not failed to remark the distinguished service performed by Major-General Sir

Colin Campbell. Her Majesty has especially noticed the brilliant conduct of the Division of Cavalry under the command of Lieut.-General the Earl of Lucan, and is deeply sensible of the gallant services of the Earl of Cardigan, and Brigadier-General Scarlett, who commanded the two brigades of Cavalry, and so nobly sustained the honour of that distinguished and important arm of Her Majesty's service.

"The conduct of the 93rd Regiment, under the command of Colonel Ainslie, merits the greatest admiration, and materially tended to the repulse of the enemy from the position which was so vigorously assailed."

"I have, etc.,

(Signed) "NEWCASTLE.

"General the LORD RAGLAN, G.C.B."

Upon this day also, the camp of the cavalry was shifted to a sheltered valley near the village of Radik, about 2 miles from Balaclava.

By a War Office letter of the 21st of December this year, a saddler sergeant was added to the strength of the regiment, from the 24th of November previous.

The regiment now commenced the construction of stabling for the horses, which was finished in February, 1855, and towards the end of December warm clothing consisting of jerseys, woollen drawers, socks, and subsequently fur coats and Turkish boots, were issued, together with clothing for the horses.

By a General Order of the 24th of December, 1854, a

medal was granted to the army in the Crimea with clasps, for the Alma, Balaclava, and Inkerman.

The regiment was now constantly employed in commissariat duties, taking supplies of biscuit to the heights for the infantry, and conveying the sick from thence to Balaclava, during all which period the health of the men, owing to continued exposure to the cold and wet, and the want of fresh meat and vegetables, was greatly affected, the prevailing diseases being diarrhœa, and scorbutic affections. The horses also suffered greatly, of which 31 were lost during the months of December, 1854, and January, 1855.

On the 18th of February, 1855, Lieutenant-General the Earl of Lucan being recalled to England was succeeded in the command of the cavalry in the Crimea by Major-General the Honourable J. Y. Scarlett, who was replaced in that of the Heavy Brigade by Colonel Hodge, 4th R. I. Dragoon Guards.

On the 19th the Cavalry Division with the Highland Brigade, under the command of Major-General Sir Colin Campbell, turned out at midnight to act in concert with a body of French troops in an attack upon the village of Tchergaum. In consequence of the severity of the weather, however, the French abandoned the project, but the messenger for the information of Sir Colin Campbell having lost his way in the darkness of the night, the British troops remained under arms till the next morning. The weather was in truth especially inclement, and occasioned many cases of frost-bite.

On the 18th of May Marshal Canrobert resigned the command of the French army to Marshal Pellisier. On the 28th of June, worn out with anxiety, suffering, and far advanced in life, Field Marshal Lord Raglan died of cholera, and was succeeded in command of the British Army by another Waterloo veteran, General Simpson.

On the 30th, Captain Davenport, Lieutenant Fitz-gerald, 68 men, and 120 horses, joined from England at Radikoi, and on the 16th and 31st of July, Captain Ainslie, with 85 men and 111 horses, arrived also from England, and a further detachment of 29 men and 35 horses, with Lieutenant Hartopp, on the 14th of August joined the regiment.

Early in the year Sardinia had entered into the coalition, and on the 16th of August, the Russians driving in the French and Sardinian outposts and crossing the Tchernaya, attacked their positions, when the cavalry division, concentrated by Major-General the Honourable J. Y. Scarlett, stood in reserve, but were not engaged during the action.

At the final storm and capture of Sevastopol on the 8th of September, the Royal Dragoons furnished a squadron to prevent stragglers from entering the city.

On the 11th of the month, Lieutenant Coleman, Cornet Cutler, with 54 men joined at Radikoi, where, on the 21st of October, an additional 40 men arrived.

The Royal regiment of Dragoons embarked at Balaclava on board the *Golden Fleece,* on the 15th of November, and landed at Scutari, where they encamped at Parda Pasha, being joined on the 19th by Cornets Curtis and Graham with 37 men from England.

In the course of the month General Simpson resigned the command of the army to Lieutenant-General Sir William Codrington, K.C.B.

On the 6th of December the regiment moved into the Zinc barracks, where it passed the winter.

In compliance with an order received, dated "Horse Guards, 1st April, 1855," the dress of the officers and men was established according to the following description:—viz., officers to wear a scarlet tunic, a white metal helmet with gilt ornaments and black plume, and a silver pouch box; the men to wear a scarlet tunic, and white metal helmet with brass ornaments and black plume.

"HORSE GUARDS, 16 *October*, 1855.

"GENERAL ORDER.

"By the Right Honble. FIELD MARSHAL VISCOUNT HARDINGE.

"The Queen has been graciously pleased to command that in commemoration of the gallant conduct of the regiment, the words 'Balaclava' and 'Sevastopol' be borne on the standards.

(Signed) "G. A. WETHERALL,
"*Adjt.-General.*"

The Crimean War was now virtually at an end, and on the 30th of March, 1856, a treaty of peace was signed at Paris.

On the 6th of April His Imperial Majesty, the Sultan, passed in review the troops from Scutari and Rubelia, when the following district order was published by Major-General Storks:—

"ASSIST. ADJT.-GENERAL'S OFFICE, SCUTARI,
"8 *April*, 1856.

"BY MAJOR-GENERAL STORKS.

"The Major-General commanding has received the commands of the Sultan to express to the officers and troops, who were reviewed by His Imperial Highness yesterday, the satisfaction His Majesty experienced at their appearance and discipline. His Imperial Majesty expressed himself in the most gracious terms, and several times desired that the troops might be informed of his approbation.

"By order,
(Signed) "C. R. ST. CLAIR,
"*Captain D.A.A.G.*"

The Royals embarked at Scutari for England on the 13th of May, having previously cast and sold by auction 48 horses, and made over 26 to the Turkish Government. They landed at Portsmouth on the 29th of the month, and marched to Aldershot, where they were inspected by Major-General the Earl of Cardigan, K.C.B., Inspecting General of Cavalry.

CHAPTER VIII.

HOME.

On the 17th of June, 1856, Her Majesty Queen Victoria came to Aldershot and inspected The Royal Regiment of Dragoons. Attended by Lieutenant-Colonel Wardlaw, the Queen walked through the temporary stables. She spoke to all the men who had been wounded, and to those who wore medals, and was pleased to express herself highly satisfied with the appearance of the regiment. On the 26th of the month, four troops proceeded by rail to Southampton, and embarked for Dublin, *en route* for Newbridge, where they arrived on the 4th of July, and found there the depot, which, having left Canterbury on the 5th of June, embarked at Liverpool on the 30th, landed at Dublin the 1st of July, and the same day continued to Newbridge.

The head-quarters, leaving Aldershot on the 7th of July, embarked at Woolwich on the 11th in the *Jura* for Dublin, arriving at Kingstown on the 13th, where landing on the following day, they marched by Lucan and Celbridge to Newbridge, arriving there on the 15th.

On the 20th of August, two troops were ordered, one to Athy, the other to Kildare, the troop at Athy moving on the 3rd of September to Kilcullen.

The regiment was inspected on the 9th of October by Major-General the Earl of Cardigan, K.B., and on the 24th by Major-General Parlby, commanding the cavalry brigade in Dublin. On the 26th, the troops at Kildare and Killallen rejoined head-quarters at Newbridge.

On the 18th of November, the establishment was reduced to six troops, strength as follows:—

Sergeants.	Trumpeters.	Farriers.	Corporals.	Privates.	Horses.
26	7	6	18	384	300

A quarter-master sergeant was at the same time added to the regiment.

In consequence of this reduction, 38 men volunteered to the Military Train, 41 to the 2nd Dragoon Guards, and 167 were discharged.

On the 10th of March, 1857, Lieutenant-Colonel Wardlaw succeeded Lieutenant-Colonel Yorke in command of the regiment, which, on the 24th, left Newbridge for Dublin; where, on arrival, leaving the staff, band, and dismounted men, the six troops of the Royals proceeded in aid of the civil power into the counties of Mayo and Leitrim, remaining there during the Elections, and returning to Dublin on the 23rd, 25th, and 29th of April, when the following letter was published in general orders:—

"GENERAL ORDER.

"ADJT.-GENERAL'S OFFICE, DUBLIN,
"29th April, 1857.

"The General commanding has much satisfaction in informing the troops who have lately been employed in aid of the civil power during the recent general elections, that he has received from the Sheriff of the County of Leitrim, and the civil authorities, several communications expressing their personal acknowledgement to Lieut.-Colonel Wardlaw of the Royal Dragoons, and the officers employed in the Counties of Leitrim and Galway, for the exertions made by them for the preservation of the peace, as well as for the firmness and forbearance which they have displayed in the discharge of a very difficult duty.

"By command of the General commanding
(Signed) "R. B. WOOD, D.A.G."

Owing to the mutiny of the Bengal Native Army, several corps of cavalry were ordered to India, and during the months of August and September the Royal Dragoons gave 1 volunteer to the King's Dragoon Guards, 41 to the 7th Dragoon Guards, and 20 to the 7th Hussars.

On the 10th of September, the establishment was increased to 8 troops, as follows:

1 colonel, 1 lieutenant-colonel, 1 major, 8 captains, 8 lieutenants, 8 cornets, 1 paymaster, 1 adjutant, 1 riding master, 1 quarter-master, 1 surgeon, 1 assistant surgeon, 1 veterinary surgeon, 1 regiment sergeant-major, 8 troop

sergeant-majors, 1 quarter-master sergeant, 50 sergeants, 9 trumpeters, 32 corporals, 8 farriers, 537 privates, and horses.

The regiment was inspected by Major-General the Earl of Cardigan on the 29th of September, and by Major-General Parlby on the 5th of October.

On the 6th of May, 1858, the regiment was inspected by Major-General Parlby, and by Major-General the Earl of Cardigan on the 9th of August.

On the formation of a 4th squadron of the regiment, application had been made for an additional guidon, when the following memorandum was received:—

"HORSE GUARDS, S.W.
"*18th August*, 1858.

"Her Majesty has been pleased to approve on the 7th inst. that regiments of Dragoon Guards and Dragoons henceforth carry but one standard or guidon; that the 2nd, 3rd, and 4th standards or guidons at present in use be discontinued, and that the authorized badges, devices, distinctions and mottoes be in future borne on what is now called the Royal, or first standard or guidon in the Dragoon Guards and Dragoons.

(Signed) "T. TROUBRIDGE,
"*Deputy Adjutant-General.*"

Major-General Parlby inspected the regiment on the 7th of October.

On the 6th of November 1 troop marched to Belfast.

In the course of this month the cloaks of the heavy cavalry were changed from scarlet to blue.

The Royal Dragoons, on the 26th of April, 1859, were despatched in aid of the civil power into the Counties of Mayo, Galway, Meath, Limerick, Antrim, and Roscommon, returning to Dublin on the 27th of May.

On the 20th of June, the regiment marched to Blessington, County Wicklow, encamped there for the night, and the next day proceeded to the Curragh of Kildare, where it encamped in a valley on the north-east side of the huts, and here the men built up walls of turf sods, which very much protected the horses in wet and stormy weather. The Royals remained 15 weeks under canvas, and, although during the last six the weather was very tempestuous, both men and horses enjoyed excellent health, nor were there any casualties.

On the 3rd of October appeared the following division order:—

"ADJUTANT-GENERAL'S OFFICE, CURRAGH CAMP,
"*3rd October*, 1859.

"DIVISION ORDER.

"The time being at hand when the troops named in the margin will be required to break camp and disperse to their several destinations, the Major-General commanding the division considers it due to them that he should record in orders his unqualified approbation of their conduct during the term that they have been under canvas, of the admirable manner with which they have borne with the comparative discomforts of camp-life, of the invariable cheerfulness under all circumstances of trial, and of their scrupulous care of their horses, whose goodly appearance, even at this moment, testifies to their unremitting attention.

"It is by such proofs as these that the character and condition of the troops may be judged; and it is with great pleasure that the Major-General acknowledges in his brief address the good conduct of the troops to which he has alluded.

"By order of Major-General E. F. Gascoigne.

(Signed) "T. C. BLANE,

"*Acting Adjutant-General.*

"C Battery R.H.A.—LIEUT.-COLONEL STRANGE.
"G Battery R.H.A.—MAJOR NEWTON.
"Royal Dragoons—COLONEL WARDLAW."

The camp accordingly broke up on the 4th of October, when 1 troop of the Royal Dragoons marched to Belfast, and a squadron to Dundalk. The head-quarters marched into the Royal Barracks, Dublin, with four troops on the 7th.

On the 12th of the month the regiment was inspected by Major-General the Earl of Cardigan, and on the 17th by Major-General Parlby.

Major-General Parlby inspected the regiment on the 11th of May, 1860. On the 8th of June the troop from Belfast joining those at Dundalk, the 4 troops joined head-quarters in Dublin on the 22nd of the month.

On the 23rd and 24th of July the regiment was inspected by Major-General J. Laurenson, Inspecting General of Cavalry, from whom it received the highest praise.

A troop was detached on the 31st to Sligo. On the

7th of August, Sergeant Edward Jewhurst was presented on parade with a silver medal and a gratuity of five pounds, he having served upwards of 25 years in the regiment. On the 18th the troop rejoined from Sligo, and on the 23rd a troop proceeded to the Camp at the Curragh, 2 troops on the same day, with a detachment of 26 men, marching to Newbridge Barracks.

In the month of September rifled carbines of the Indian pattern were issued to the regiment, which was inspected by Major-General Parlby on the 9th of October.

On the 19th of March, 1861, the Royals were inspected by Major-General Parlby, and on the 30th of May the head-quarters left Dublin for Newbridge.

On the 12th of June, the establishment was reduced from 626 men and 438 horses to 600 men and 400 horses.

On the 13th of August, General H.R.H. the Duke of Cambridge, K.G., made an inspection of the barrack-rooms, hospital, and stables of the regiment at Newbridge, and was pleased to compliment Colonel Wardlaw upon the cleanliness of the rooms and the excellent condition of the horses.

On the same day His Royal Highness inspected the Cavalry Brigade on the Curragh, under the command of Major-General Parlby, composed of the following regiments:—The Royal Dragoons, 3rd K. O. Hussars, 4th Q. O. Hussars, 5th Lancers, 11th Prince Albert's Hussars, and the 14th K. Hussars.

The following is an extract from the Brigade Orders issued upon this occasion:—

"CAVALRY BRIGADE OFFICE, NEWBRIDGE,
18*th Aug.*, 1861.

"Major-General Parlby congratulates the Brigade on his having received the commands of H.R.H. the Duke of Cambridge to express to the troops his perfect approval of the admirable manner in which the Cavalry Brigade manœuvred when under the personal command of His Royal Highness.

(Signed) "W. PARLBY, Major-General,
"*Army Cavalry Brigade.*"

The regiment being ordered to furnish a guard of honour for Her Majesty the Queen during her visit to Killarney, on the 17th of the month a strong detachment left Newbridge, *en route* to Killarney, under the command of Captain Graham, and rejoined head-quarters on the 7th of September.

The Royal regiment of Dragoons formed part of the division reviewed by Her Majesty Queen Victoria on the 24th of August, on the Curragh of Kildare, when H.R.H. Prince Albert personally complimented Colonel Wardlaw upon the general appearance of the regiment.

On the 31st of the month Major-General Parlby issued the following order on relinquishing the command of the cavalry brigade in Dublin:—

"CAVALRY BRIGADE OFFICE, NEWBRIDGE,
"31*st August*, 1861.

"On resigning the command, Major-General Parlby cordially thanks the officers and men of the Cavalry Brigade for the very zealous and willing assistance they have invariably afforded him in the execution of his duties. He cannot but feel that it is owing to their exertions that he is now able to hand over the brigade to his successor in that high state of discipline, which on a recent occasion called forth in so marked a manner the approbation of His Royal Highness the General Commanding in Chief.

"In bidding farewell to the Royal Dragoons, now on the eve of departure from Ireland, after a service of 5 years under his immediate command, Major-General Parlby cannot let this opportunity pass without expressing his esteem for this noble regiment, whose exemplary conduct and soldierlike bearing reflect the highest credit on all ranks, and have won the approval of every officer under whom they have served.

(Signed) "W. PARLBY, Major-General.
"*Commanding Cavalry Brigade.*"

The regiment was now ordered to Dublin, there to embark for England as follows, viz :—The 1st squadron on the 3rd of September, the 2nd on the 9th, and the 3rd on the 13th ; the head-quarters on the 18th, and landing at Liverpool, it proceeded 3 troops to Sheffield, 2 to Coventry, and head-quarters with 3 troops to Birmingham.

Major-General Laurenson made his inspection of the regiment on the 18th of October.

By a War Office letter of the 29th of November, a sergeant-instructor of musketry was added to the strength of the regiment, from the 5th of the month.

From this date nothing beyond ordinary changes of quarters took place until on the 9th of July, 1862, the regiment was inspected by Major-General Laurenson, and again on the 23rd of September following.

The regiment was now ordered to proceed to Aldershot as follows :—The detachments from Coventry on the 7th and 11th of May, 1863, from Weedon on the 11th and 12th, and head-quarters from Birmingham on the 15th ; the whole concentrating in the C and F Lines, North Camp, Aldershot, on the 22nd. On the 28th of the month it was inspected by Major-General Hodge, C.B., commanding the cavalry brigade.

On the 30th of June the regiment went under canvas on Cove Common until the 8th of August, when it formed part of a column proceeding to Sandhurst, whence it returned on the 12th to Aldershot, and was quartered in the East Cavalry Barracks.

Major-General Laurenson inspected the regiment on the 21st of September, and Major-General Hodge, C.B., on the 2nd of October.

A detachment under Captain Robertson proceeded on the 3rd of November to Guildford, to assist the magistrates, returning on the 19th.

Major-General Hodge, C.B., made his inspection of the Royal Dragoons on the 2nd of May, 1864, and on the 14th of July, the regiment accompanied a column under Major-General Brook Taylor to Wolmer, encamping there until the 19th, when it returned to Aldershot.

The regiment was inspected by Major-General Hodge, C.B., on the 3rd of October, and by Major-General Laurenson on the 8th and 9th ensuing.

On the 3rd of May, 1865, the Royals were inspected by Major-General Hodge, C.B., and left Aldershot on the 7th and 8th of August, 3 troops to Shorncliffe on the former day, and head-quarters and 4 troops on the latter to Brighton.

Major-General Lord George Paget, C,B., Inspecting-General of Cavalry, inspected the regiment on the 20th of October.

The detachment at Shorncliffe under Major Ainslie joined the head-quarters on the 31st of March, 1866, a squadron being billeted at Lewes, and the remainder putting up in the Barracks at Brighton, where on the 2nd of April the entire regiment assisted at a Review of the Volunteers, commanded by Lieutenant-General Sir R. Garrett, K.C.B., K.H., at which were present upwards of 20,000 men.

On the 3rd, the regiment was inspected by Major-General Lord Paget, C.B., and on the 4th, 12th, and 13th of the month, it left Brighton *en route* for Manchester, detaching troops to Ashton-under-Lyne, Bury, and Burnley. On the 18th of May, a troop marched to Nottingham in aid of the magistrates there, but returned on the 2nd of June. On the 10th of January, 1867, the Royal Dragoons were inspected by Major-General Sir John Garvock, K.C.B., commanding the Northern District, and on the 9th, 10th, 11th, and 16th of April they embarked at Liverpool for Dublin, whence on arrival they marched to Newbridge.

On the 23rd and 24th of the month, they were

inspected by Brigadier-General H. D. White, C.B., commanding the Cavalry Brigade in Dublin.

Colonel Robert Wardlaw, retiring on half-pay on the 8th of May, was succeeded in the command of the regiment by Major James Ainslie.

On the 11th of July, 4 troops, under the command of Major William Coney, marched to the assistance of the civil power to Dundalk, whence returning on the 19th they proceeded to the Curragh where they encamped, and were joined by 3 troops from head-quarters.

The regiment was inspected on the 20th of August by Major-General Lord George Paget, C.B., and on the 25th and 26th of September by Brigadier-General H. D. White, C.B.

On the breaking up of the camp at the Curragh for the winter, the Royals moved as follows: head-quarters and two troops to Longford on the 11th and 14th of October, on the 11th also one troop to Castlebar and one to Athlone, on the 24th a squadron under Major Coney to Limerick, and at the same time two troops moved into the barracks at Newbridge.

On the 5th of February, 1868, one troop left Newbridge to relieve a troop of the 9th Lancers at Waterford. On the occasion of the trial of Fenian prisoners at Sligo, a troop was directed to be sent from Longford to remain at that place during the assizes there, which arrived accordingly at Sligo on the 26th of February, returning to Longford on the 6th of March. On the

13th of April, a troop left the barracks at Newbridge for the camp at the Curragh, there to relieve a troop of the 10th Hussars on vedette duty at the rifle ranges, and on the 14th of May the troop at Waterford left that place for Newbridge, arriving on the 18th.

The regiment was inspected by Brigadier-General Archibald Little, C.B., commanding the Cavalry Brigade in Dublin, on the 14th and 15th of May.

On the 22nd of July came an order for the Royal Dragoons to join the camp at the Curragh, there to remain during the drill season. In consequence of this on the 28th and 29th of the month, leaving a troop at Limerick and 1 at Athlone, the head-quarters with 2 troops leaving Longford, reached the Donelly Hollow, Curragh, where they encamped. A troop from Limerick marched into Newbridge barracks on the 1st of August, on which day the troop on vedette duty being relieved by one of the Carabineers joined head-quarters, which on the 4th were further increased by the arrival of the troop from Castlebar.

A new kind of carbine, the "Snider breechloader," having been approved for the cavalry, 520 of these arms had been received on the 28th of March previous, but were not issued to the regiment until the month of August, in consequence of the old pattern carbine buckets not being available for these new arms; and therefore pending the issue of new buckets the regiment at mounted parade carried no carbines.

Major-General Lord George Paget, C.B., inspected the regiment on the 7th of September.

A battery of the Royal Artillery requiring accommodation in the temporary stables at Donelly Hollow, a troop of the Royals moved into the barracks at Newbridge on the 16th.

The regiment was inspected by Brigadier-General A. Little, C.B., commanding the Cavalry Brigade, Dublin, on the 13th of October, and on the breaking up of the camp at the Curragh, the Royal Dragoons moved to Dublin, 2 troops on the 14th from Newbridge, head-quarters and 3 troops from the Curragh, and 1 troop from Newbridge on the 17th; the troops from Athlone and Limerick on the 21st arriving in Dublin on the 24th and 27th, when the entire regiment concentrated in the barracks at Island Bridge.

Early in November the services of the Royals were required during the ensuing general elections, and on the 7th of the month a squadron marched to Limerick.

Two troops on the 9th respectively to Galway and Philipstown, and from the latter on the 28th to Boyle. A detachment went to Balbriggan from the 20th to the 22nd, and on the 21st a troop left Dublin for Boyle, joining the troop already stationed there, until on the 8th of December, the squadron returned to Dublin.

A troop was sent to Wicklow on the 23rd, returning on the 27th, and on the 23rd also a troop proceeded to Trim, returning on the 2nd of December. The squadron from Limerick, having left on the 10th of December, arrived in Dublin on the 15th. Early in January, 1869, new carbine buckets, suitable for carrying the breech-loading arms, were supplied, when the swivels being no longer required were given into store, and the pouch belts were altered to the pattern worn by the Lancers.

On the 7th of March, died, at the age of 97, General Sir Arthur B. Clifton, G.C.B., Colonel of the regiment, with which he had so long been associated with such distinction, and was succeeded on the 8th of the month by Major-General Charles P. de Ainslie.

A General Order, dated "Horse Guards, 13th of April, 1869," directing that the organisation by squadrons should be adopted throughout the cavalry, the 8 troops of the Royal Regiment of Dragoons were in consequence formed into 4 squadrons, commanded respectively by Captains Coleman, Graham, Radford, and Hall; while at the same time the establishment of the regiment was constituted as follows: 1 colonel, 1 lieutenant-colonel, 1 major, 8 captains, 8 lieutenants, 11 cornets, 1 paymaster, 1 adjutant, 1 ridingmaster, 1 quartermaster, 1 surgeon, 1 assistant-surgeon, 1 veterinary surgeon, 1 regimental sergeant-major, 1 trained bandmaster, 1 regimental quartermaster-sergeant, 4 squadron sergeant-majors, 4 quartermaster-sergeants, 1 paymaster-sergeant, 1 armourer, 1 farrier major, 4 farriers, 1 sadler-sergeant, 1 hospital-sergeant, 1 orderly-room clerk, 1 sergeant instructor of musketry, 1 sergeant cook, 24 sergeants, 1 trumpet-major, 8 trumpeters, 24 corporals, 12 shoeing smiths, 4 saddlers,

1 saddle-tree maker, 426 privates, 523 inclusive of officers, 344 troop horses.

On the 19th and 20th of May, the regiment was inspected by Major-General Sir A. Little, K.C.B., and in the month of July it was employed in the North of Ireland, during the Orange celebration of the 12th at Limerick, Newry, Banbridge, and Armagh, part of the corps remaining at Dundalk under Major Coney.

On the 11th and 13th of this month, Major-General Lord George Paget, C.B., inspected the regiment, and expressed his entire satisfaction with its appearance, drill, and the condition of the horses.

Troops being required to assist the civil power during the elections in the County Antrim, a squadron under Major Coney proceeded by rail on the 17th to Belfast, and a second to Lisburn on the 18th of August, returning to Dublin on the 23rd and 25th of the month.

The regiment was inspected by Brigadier-General R. Wardlaw, C.B., on the 1st of October, about the middle of which month it received orders to move to Cahir and out stations, there to relieve the Scots Greys; and in consequence, on the 22nd of the month, a squadron left Dublin, one troop for Ballincollig and a detachment for Waterford. The 2nd squadron, under Major Coney, marched on the 23rd and 25th for Cork, arriving on the 2nd and 3rd of November. On the 23rd, the 4th squadron, and on the following day the head-quarters and 3rd squadron, left Dublin for Cahir, arriving there on the 29th and 30th respectively. A small detachment under Lieutenant the Honourable C. G. Trench left Dublin on the 26th of October for Clonmel, arriving there on the 1st of November. The dismounted party went by rail to Cahir on the 29th of October.

In the course of November, detachments of the Royal Dragoons were employed at Tipperary, Cahir, and Clonmel, on the occasion of the county elections; and owing to the unsettled state of the country, the King's Dragoon Guards being brought over to augment the cavalry in Ireland and ordered to Cahir, the Royals were in consequence moved as follows:—viz., head-quarters to Ballincollig, one squadron to Limerick, half a squadron to Fermoy, and half a squadron to Ennis.

On the 23rd of January, 1870, half of the 2nd squadron at Cork marched to Mallow for the election for that borough, returning on the 2nd of February.

On their return from India, the 2nd Dragoon Guards, "Queen's Bays," were mounted by transfers of horses from the regiments of cavalry at home, and the Royal Dragoons furnished a quota of 44 bay horses, which, being marched from their several stations to Dublin, were there given over to a party of the Bays on the 15th of March. During this month, orders were received for the cavalry to revert to the former troop organisation from the 1st of April, when the regiments were directed to be formed into 7 troops instead of 8 as previously, and accordingly 7 troops were constituted as nearly as possible as they had originally existed, one troop being distributed among the remainder, making a total of 458 men and 300 horses. These arrangements leaving but one troop at head-quarters, the troop from Ennis was moved to Limerick, whence a troop was brought in to Ballincollig, which changes were carried out on the 21st of March.

Major - General Campbell, C.B., commanding the South-Western District, inspected the regiment on the 24th of May.

On the 15th of August, the establishment was increased by 83 rank and file, and 50 horses.

The regiment was inspected by Major-General R. Wardlaw, C.B., on the 7th of October, and during the month some changes of troop quarters were made.

The establishment was increased on the 1st of February, 1871, by an additional troop, making a total of 28 officers, 607 non-commissioned officers and men, and 384 troop horses.

One troop from Limerick marched to Gort on the 18th for the County Galway elections, returning on the 23rd; and on the 28th, a troop left Ballincollig for Dunmanway, but returned on the 1st of March. On the 10th, a troop proceeded to Bandon; and on the 21st, a detachment of 75 men and horses moved into Cork; whence, on the 27th, a troop marched to Fermoy, there to be stationed. On the 3rd of the month, orders were received for the regiment to be held in readiness to move to the Curragh Camp, recruits and young horses to the Royal Barracks, Dublin.

In the course of the month of April helmets of a new pattern were issued to the regiment.

The Royal Regiment of Dragoons now moved to the Curragh Camp as follows:—the squadron from Cork, under Major Coney, on the 5th of May; the troops from Fermoy on the 8th of June; from Ballincollig one troop on the 9th; head-quarters, under Lieutenant-Colonel Ainslie, from Ballincollig on the 12th; troop from Limerick on the 13th; and the troop from Bandon on the 13th. On the 11th, the young horses marched from Ballincollig, the recruits proceeding by rail the same day; and on the 13th, the dismounted men of head-quarters proceeded also by rail to the Royal Barracks, Dublin; and on the 24th of May, the regiment had assembled at the Curragh, and was quartered in huts in K. square. The regiment was inspected by Major-General R. Wardlaw, C.B., on the 5th of May.

On the 24th of July the recruits and young horses left Dublin for Dundalk; the dismounted men by rail on the 26th.

On the 2nd of October the head-quarters of the Royal Dragoons commenced their march from the Curragh to Dundalk, arriving there on the 5th, 6th, 7th, and 9th; the troop for Belturbet on the 11th, Belfast on the 14th, and on the 18th the regiment was inspected by Major-General Wardlaw, C.B.

The Major-General inspected the regiment on the 12th of April, 1872, in the course of which month pantaloons and long boots were issued in place of the leathered overalls and Wellington boots hitherto worn.

On the 9th of May, agreeably to orders received, the

Royal Dragoons moved to Dublin and the Royal Barracks in the following order, viz.: on the 9th a squadron under Major Graham; a second squadron on the 13th, and on the same day the Riding-master and Veterinary Surgeon with the young horses; and on the 16th the head-quarters and 1 squadron under Lieutenant-Colonel Ainslie. The troops from Belturbet and Belfast marched respectively on the 15th and 16th, the dismounted men, women, and children by rail on the 17th of May.

The Royal Regiment of Dragoons was inspected on the 17th of April, 1873, by Major-General Sir Thomas Steele, K.C.B., commanding the Dublin District, who expressed great satisfaction with its appearance, and was also pleased to remark that its conduct while under his command had been most creditable, which he would not fail to bring to the notice of H.R.H. the Field Marshal Commanding in Chief.

A new guidon was issued to the regiment during this present month of April.

Notice having been received of the proposed move of the Royal Dragoons to Scotland, the *Duke of Leinster* steamer was chartered to convey the whole regiment from Dublin to Glasgow, and on the 24th of April 2 troops embarked at the North Wall, and on landing at Glasgow one of these troops proceeded to Hamilton, the other to Piershill Barracks, Edinburgh. On the 28th a second squadron embarked, and, landing at Glasgow, proceeded to Edinburgh. Three troops, under the command of Major Graham, embarked for Glasgow on the 1st of May, and thence marched to Hamilton; and on the 5th, 1 troop with head-quarters, under Lieutenant-Colonel Ainslie, embarked for the same destination, and thence moved to Edinburgh. On the 19th 1 troop was brought in from Hamilton, and on the 19th of June a second troop was brought in from Hamilton to Edinburgh.

Major-General Sir Thomas McMahon, Bart., C.B., Inspector-General of Cavalry in Great Britain, inspected the regiment on the 11th, 12th, and 13th of August.

On the occasion of the marriage of H.R.H. the Duke of Edinburgh, K.G., on the 23rd of January, 1874, a portion of the Royals was employed on patrol duty in the streets of Edinburgh during the illuminations, when Major-General Sir John Douglas, K.C.B., commanding the North British District, was pleased to express his approval of the conduct of the regiment as having been most orderly and soldierlike.

In the course of the months of February and July some changes of distribution took place, and on the 10th, 11th, and 12th of August, Major-General Sir Thomas McMahon, Bart., C.B., made his inspection of the regiment.

In 1875, some small changes of quarters were made, and towards the end of the month notice was received of a proposed move of a regiment to York.

After a field-day in honour of the Queen's birth-day on the 29th of May Major-General Sir John Douglas,

K.C.B., addressed to Colonel Ainslie the following letter, which he desired to be sent to the regiment:—

"EDINBURGH, 30*th May*, 1875.

"MY DEAR COLONEL,

"I intended to say a few words to your regiment on Saturday, but I was engaged with Lord Rosslyn when you moved off. I shall be much obliged by your reading this letter to the regiment, and telling them how much I regret the departure of the Royals from my command. During the two years the regiment has been in Edinburgh and Hamilton nothing could be better than their conduct; and their appearance, either on parade, in the field, on orderly duty, or in the streets, has always been very fine. They are a magnificent body, both of men and horses, and I am certain that, go where they may, the Royal Dragoons will never find a regiment superior to themselves.

"Believe me, very truly yours,
(Signed) "T. DOUGLAS,
"*Major-General.*"

On the 31st of May 1 troop left Edinburgh for York, and on the 1st of June 1 troop under Major Graham left Hamilton also for York; on the 2nd 1 troop from Edinburgh, whence on the following day marched the sick and young horses; on the 4th and 7th 1 troop from Edinburgh, and 1 on the latter day from Hamilton for York; on the 9th 1 troop, and on the 11th 1 troop with head-quarters, from Edinburgh for York, the dismounted men, women, and children proceeding by rail

The entire regiment concentrated at York by the 24th of June.

On the 1st and 2nd of September it was inspected by Major-General Sir Thomas McMahon, Bart., C.B.

On the 9th of April, 1876, after an illness of 6 days, Colonel James Ainslie died in York Barracks. His remains were carried by the regiment to the railway station, whence they were conveyed to Elvington, Gladsmuir, N.B., to be there interred in the family vault. A handsome memorial window has been put up in the chapel of the York Cavalry Barracks by the officers, non-commissioned officers and men who had served with him in the Royal Regiment of Dragoons. He was succeeded in the command by Major Graham.

Towards the end of June the regiment was held in readiness to move to Norwich, Ipswich and Colchester, for which latter destination one troop, with a detachment, left York on the 11th of July, on which day also the sick and young horses, with Lieutenant T. Travers Clerk and the veterinary surgeon, marched for Norwich. On the 12th Major Hutton, with one squadron, marched for Ipswich. On the 13th and 15th respectively, a squadron marched for Norwich, and on the 18th one troop and head-quarters, under Lieutenant-Colonel Graham, left York for Norwich, the distribution of the regiment being as follows :—

Norwich	5 Troops and head-quarters . . .	237	horses
Ipswich	2 „	76	„
Colchester . . .	1 Detachment	61	„

The dismounted men, women and children to Ipswich and Norwich by rail on the 18th and 19th of the month.

On the 16th of March, 1877, were received orders of readiness for Aldershot, whither the regiment moved as follows, viz.: On the 9th of April, Lieutenant-Colonel Graham, with head-quarters and two troops from Norwich, and on the same day a third troop from thence, and the squadron from Ipswich, under Major Hutton, and on the 11th and 12th two other troops from Norwich. On the 13th the dismounted men, women, and children arrived at Aldershot by rail, and on the 21st the sick and young horses, the whole regiment being quartered in the West Cavalry Barracks with the following strength:—

26 officers, 626 men of all ranks, 384 horses, and 8 troops.

The Cavalry Brigade at Aldershot at this time was commanded by Major-General R. Wardlaw, C.B., and comprised the Royal Dragoons, 8th R. I. Hussars and 17th Lancers.

On the 9th of July the regiment accompanied the Cavalry Division, consisting of the 1st Life Guards, the Royal Dragoons, 17th Lancers, 8th, 18th and 20th Hussars, under Major-General Wardlaw, to Ascot Heath, where they encamped for two nights; and on the 10th they formed part of an army corps, which, assembling in Windsor Park, marched past the Queen about 5 o'clock in the afternoon, returning to Aldershot on the

11th. Upon this occasion the regiment turned out 333 troop horses.

On the 26th and 27th of September the regiment was inspected by Major-General Wardlaw, C.B., Inspecting General of Cavalry in Great Britain.

The Martini-Henry carbines were issued on the 15th of March, 1878, when the regiment was at once put through a course of musketry.

In the month of April, owing to the state of affairs in Turkey, Her Majesty's Government deemed it expedient to call out the 1st Class Army Reserve to form a 1st Army Corps, of which the 1st, 4th and 5th Dragoon Guards, the Royal Dragoons, 8th Hussars and 17th Lancers composed the Cavalry, the Royals being at the same time augmented by 80 troop horses, 24 draught horses, and receiving 3 waggons, 1 field forge, and 1 ammunition cart. Fifty-seven of the additional horses were purchased in Ireland from T. Manley, the remainder in England, and 30 were attached for the purpose of training to the 16th Lancers. Twenty-one men of the Cavalry Reserve were attached to the Royals, of which number, nine being found medically unfit they were sent to the head-quarters of their pension districts to be discharged; the remainder were put through a course of riding school and musketry instruction, and performed their regimental duties. Peace being restored between Russia and Turkey, the Reserve had all returned to their homes by the 3rd of July.

The regiment was inspected by Major-General

Wardlaw, C.B., on the 30th of September and 1st of October.

On the 27th of February, 1879, the Royal Dragoons received orders of readiness for Colchester and Norwich, and on the 22nd of March one troop left Aldershot for Norwich, being followed on the 24th by a second for the same destination. The head-quarters and 2 troops, under Lieutenant-Colonel Graham, marched also on the 24th for Colchester, as also 3 troops under Major Hutton, and one other separately, which 6 troops reached their destination on the 28th. The dismounted men, women and children proceeded as usual by rail.

On the 2nd of June, Captain Dickson embarked upon special service for the Cape of Good Hope, and in the course of the month an exchange of troops took place between Colchester and Norwich.

Lieutenant H. A. Amyatt Burney, having volunteered for service at the Cape of Good Hope, was attached to the King's Dragoon Guards, and embarked with that corps on the 27th of February, 1879.

Major-General W. C. Radclyffe, C.B., commanding the Eastern District, inspected the regiment on the 25th and 26th of July, and on the 2nd and 3rd of October it was inspected by Major-General Sir Fred. Fitzwygram, Bart., Inspecting General of Cavalry in Great Britain, on which latter date also were received orders of readiness for Manchester, Liverpool and Birmingham on the 16th of the month; therefore 2 troops left Colchester for Liverpool on the 20th, the squadron at Norwich marched for Birmingham, and on the same day a squadron left Colchester for Manchester. The head-quarters and 2 troops, under Major Hutton, left on the 28th for Manchester. The dismounted men, women and children by rail on the 6th of November; 4 sergeants and 78 rank and file were left at Colchester under Lieutenant Travers Clerk for the purpose of assisting in the care of a number of Austro-Hungarian horses purchased for the 3rd King's Own Hussars, which party, after the return of the head-quarters of the 3rd from India, left for Manchester by rail on the 12th of December, where they arrived the same day.

Captain Dickson and Lieutenant H. Amyatt Burney arrived in England on the 5th of November from the Cape of Good Hope, where Captain Dickson, with the rank of Deputy Assistant-Adjutant General, had served as second in command of the Native Carrier Corps, under Major Schwabe, 16th Lancers, from the 8th of July, 1879, to the end of the campaign.

Lieutenant Burney was present at the capture of King Cetewayo, on the 28th of August, 1878, by the squadron of the King's Dragoon Guards, under Major Marter.

Lieutenant-General G. H. P. Willis, C.B., commanding the Northern District, inspected the Royal Dragoons at Liverpool on the 2nd of August, 1880, and the head-quarters at Manchester on the 6th following.

On the 11th, the head-quarters and two troops marched to Liverpool, embarking there for Ireland on the 12th, and landing at Dublin on the 13th, where they temporarily put up in the Royal Barracks. The Liverpool squadron, embarking on the 14th, landed on the following day at Dublin, and proceeded to the same quarters. The Birmingham squadron moved on the 13th, embarked on the 18th at Birkenhead, and, landing in Dublin on the 19th, joined the two former divisions; and the 2nd squadron from Manchester, leaving on the 16th, embarked on the 17th at Liverpool, landed on the 18th in Dublin, and moved into the Royal Barracks.

On the 16th of August the head-quarters and one troop of the regiment marched for Longford, and on the same day one troop marched to Ballinrobe and one to Castlebar. On the 17th one troop marched to Longford, on the 19th two troops proceeded to Gort and one to the Curragh Camp, there to complete its annual course of musketry, whence on the 31st of the month it moved to Carlow. On the 20th one troop left Dublin for Longford.

At this period the state of Ireland was in the highest degree unsatisfactory: political agitation, crime, and outrage prevailing in all directions, of which for the moment the most prominent instance was the murder on the 25th of September, in County Galway, of Viscount Mountmorres. The troops were consequently much dispersed and harassed.

On the 12th of October a troop moved from Longford to Athlone, there to be stationed.

On the 12th of January, 1881, Captain Perceval's troop left Gort for Longford, reaching there on the 15th, having been compelled to march the whole distance on foot, owing to the inclemency of the weather and the very bad condition of the roads.

On the 1st of April this year the non-commissioned officers and men received for the first time forage caps from the Royal Army Clothing Department, the caps forming henceforward an article of annual clothing.

By the official return of the "Establishment of the Army," issued with Army Circulars of 1st May, 1881, the regimental establishment was reduced from 601 to 469 non-commissioned officers and men, and from 379 to 300 troop horses. This order took effect from the 1st of April, 1881, and, by a return of "Changes in the Regimental Establishment of the Army 1881-82," the establishment of regimental officers was as follows, and took effect from the 1st of July, 1881, viz.: 2 lieutenant-colonels, 3 majors, 5 captains, 11 lieutenants, 1 adjutant, 1 ridingmaster, 1 quartermaster—total, 24.

The regimental sergeant-major, bandmaster, and schoolmaster were promoted to warrant officers, and this new organisation took place equally from the 1st July, 1881.

On the 30th of May one troop of the Royal Dragoons left Longford for Dublin, and the day following the head-quarters and one squadron, under Major Hutton, marched for the same destination. On the 1st of June, moved also the troops from Ballinrobe, Gort, Athlone, and Castlebar to Dublin; and the troop from Carlow, having proceeded on the 29th to the Curragh Camp for musketry instruction, rejoined in Dublin on the 8th of July, the regiment occupying the barracks at Island Bridge, with the exception of one troop in the Royal Barracks.

On the 10th of September, agreeably to instructions from the Horse Guards of the 12th of July previous, and in consequence of the reduction in the establishment, 20 men and 40 horses were transferred to the 7th P. R. Dragoon Guards.

Consequent upon a communication from the Horse Guards, dated 18th November, 1881, Her Majesty was graciously pleased to sanction the badge of an "Eagle" with No. 105 being worn on the forage caps of the officers of the Royal Dragoons, the cap in question being of the pattern approved for officers on active service and at peace manœuvres.

On the 24th of January, 1882, in consequence of the very disturbed state of Ireland, the regiment was ordered to detach 3 non-commissioned officers and 58 men upon what was called "protection duty." These men were divided into parties of from 2 to 5 each, and quartered with gentlemen whose lives had been threatened. It was their duty to escort, and never to lose sight of, these gentlemen when moving from their residences, and in most instances the ground surrounding these "protection posts" had to be carefully examined both by day and night, and the various approaches patrolled.

On the 6th of May occurred one of the most daring and atrocious murders that ever disgraced the history of any nation, for on that day were assassinated in Dublin, in the Phœnix Park, close to the Vice-Regal Lodge, and in broad daylight, Lord Frederick Cavendish, Chief Secretary for Ireland, and Mr. Burke, Under Secretary.

On the 8th of June Corporal Wallace, of the Royal Dragoons, and the gentleman he was escorting, Mr. Burke, of Rabassano Park, Cranghuell, were assassinated close to Castle Tayler, near Gort, having been shot from behind a wall. The remains of the corporal having been conveyed to Dublin were interred in Grange Gorman Cemetery, Lord Clarina, the general officer commanding the district, with his staff and a number of officers from the various corps in garrison, and a strong body of non-commissioned officers and men, attending the funeral. This murder created a very strong feeling at the time among the military.

One sergeant and 16 men were transferred on the 19th of August to the 4th R. I. Dragoon Guards for service in Egypt.

The "protection posts" in some cases becoming unnecessary, and in others the dragoons being relieved by infantry, those of the Royals rejoined their regiment on the 8th of September. Upon this arduous and harassing duty they had been very vigilant, and on their return they were complimented by his Excellency Earl Spencer, Lieutenant-General and General Governor of Ireland, and in district orders by Lieutenant-General Sir Thomas Steele, K.C.B., Commander of the Forces.

On the 20th of September of this year, Her Majesty was graciously pleased to approve of the Royal Regiment of Dragoons bearing on its guidon the word "Dettingen," in commemoration of its conduct in the battle fought there on the 27th of July, 1743.

One sergeant and 14 men were transferred on the 14th of October to the 4th R.I. Dragoon Guards for service in Egypt.

On the 22nd of February, 1883, instructions were received from the War Office that the establishment of the Royal Regiment of Dragoons was to be raised from 300 to 400 troop horses, which additional number were to be purchased before the 31st of March ensuing, up to which date, however, only 80 had been procured, owing to the shortness of the time allowed and the scarcity of horses at that time of the year.

On the 10th of July, one troop under Major Maclean was ordered to Navan, and on the 30th of August, a troop moved into Island Bridge Barracks from Portobello.

After divine service on Sunday, the 20th of September, at the Royal Hospital, the regiment, being formed up, was addressed by General Sir Thomas Steele, K.C.B., Commanding the Forces in Ireland, who said "it would be a pleasing duty to him in forwarding the Confidential

Report of the regiment to H.R.H. the Field Marshal, Commanding in Chief, to report the very high state of efficiency and discipline of the Royal Dragoons. He had had the regiment under his command on three occasions, and during the last two and a half years it was under his personal directions. That the absence of crime, and the appearance of the non-commissioned officers and men, both on and off duty, reflected the greatest credit on the regiment."

The general concluded his remarks by wishing the officers, non-commissioned officers, and men "Good-bye."

A troop marched on the 2nd of October for Athlone. On the 14th, three troops under Major Dickson marched to Newbridge, and the day following head-quarters and 3 troops under Lieutenant-Colonel Hutton marched for the same destination, a troop at the same time proceeding to the Curragh Camp on vedette duty.

On the 14th of January, 1884, Lieutenant Rhodes proceeded on service to Egypt, where he was posted to the corps of Gendarmerie.

On the 1st of April, the troop from the Curragh moved to Athlone, being relieved by one from Newbridge, where on the 5th the troop arrived which had been relieved at Athlone. On the 5th of May, Major Maclean with the troop at Navan left for Newbridge, arriving on the day following.

Major Dickson embarked on the 18th of September for Egypt, in charge of boats despatched with the expedition sent for the relief of Major-General Gordon at Khartoum, and on the 19th Brevet-Major Gough, who on the 27th of February had exchanged with Captain Amyatt Burney from the 52nd Light Infantry, Lieutenant Burn Murdoch, 2 sergeants, 2 corporals, 1 trumpeter, and 38 privates, proceeded to the like destination as the contingent ordered for the Camel Corps, forming for the Nile expedition, of which number 2 privates died at Assouan, and Major Gough with 13 men was killed at the battle of Abu Klea, on the 17th of January, 1885.

On the 20th of November, 1 sergeant and 2 men embarked for Beloochistan as signallers in the expedition under Sir Charles Warren in South Africa, where they were attached to the Volunteer Horse of Colonel the Honourable P. Methuen, C.B., Scots Guards.

On the 28th of January, 1885, Major Tidswell embarked for Egypt for duty with the Camel Corps, *vice* Gough.

On the 20th of February, a squadron commanded by Lieutenant-Colonel Russell left Newbridge for Belfast, arriving on the 26th, whence on the 14th of March a troop was detached in aid of the civil power to Londonderry, returning on the 20th.

On the 14th of March, a troop proceeded from Belfast to Londonderry, arriving there on the 16th, in aid of the civil power, and on the 21st returned to Belfast.

On the 16th of April, Captain T. B. Nichols' troop

left Athlone for Londonderry, arriving there on the 24th, in order to form an escort for the Prince and Princess of Wales during their stay in that city, whence on the 27th it marched for Dundalk, arriving on the 30th. On the 21st of the month, Captain O'Shaughnessy with Lieutenant Shaw, 1 trumpeter, and 36 non-commissioned officers and men, left Newbridge for the purpose of escorting the Prince and Princess of Wales from the railway station at Naas to the race-course at Punchestown, on the occasion of their Royal Highnesses' visit to the races. A similar escort under Captain Nichols and Lieutenant Mesham attended His Excellency the Lord Lieutenant of Ireland, who accompanied the Royal party. These escorts returned to Newbridge the same evening, and on the second day of the races, in consequence of the inclemency of the weather, the escorts were ordered to be reduced to the strength of the ordinary travelling escorts. On the second day, therefore, the escort of H.R.H the Prince of Wales was commanded by Lieutenant Shaw, and that of the Lord Lieutenant by Lieutenant Browne.

On the 19th of May Colonel Hutton retired upon half pay, and was succeeded on the day following by Lieutenant-Colonel F. S. Russell.

On the 20th of the month two troops left Newbridge for Dundalk, there to be stationed, and arrived on the 22nd. On the 24th a detachment of young and sick horses, under the ridingmaster, left for the same destination, arriving on the 28th. A second party similarly composed, under Lieutenant Carr Ellison, marched on the 26th for Dundalk, arriving on the 30th ; and on the 26th also, 3 troops under Major Maclean marched for Dundalk, where they arrived on the 28th. The dismounted men, with the women and children, under Lieutenant and Adjutant Robertson, left on the 30th of the month by rail, and arrived at Dundalk the same day.

On the 18th of June, Major Tidswell, on his way home in command of the detachment of the Heavy Camel Corps in Egypt, died from diarrhœa at Abu Limber.

Captain Burn Murdoch and 21 rank and file, the remnant of the detachment of the regiment which had formed part of the Heavy Camel Corps, disembarked on the 15th of July from the steamship *Australia* at Cowes, and having been inspected by Her Majesty the Queen, proceeded to Dundalk, where, on arriving on the morning of the 17th, the regiment was drawn up to receive them, and in the evening they were entertained at a dinner provided by the officers.

On the 14th of August Captain O'Shaughnessy, Lieutenant Sclater-Booth, and 40 non-commissioned officers and men, proceeded by special train from Belfast to Moneymore in aid of the civil power, and returned by march route to Belfast on the 18th of September.

On the 15th of September, Captain Rhodes returned to regimental duty after his meritorious services in Egypt, in which country he had served with the expedition into the Soudan for the relief of Tokar, and was present at the affair of El Teb on the 29th of February, and Talmath 13th of March, 1884, as aide-de-camp to Brigadier-General Sir Herbert Stewart, Commanding the Cavalry. He subsequently served in the same capacity with Sir Herbert Stewart, during his command of the Cavalry and Camel Corps with the expedition under Lieutenant-General Lord Wolseley for the relief of Major-General Gordon, and was present in the actions at Abu Klea on the 17th, and Gabal on the 19th of January, 1885, in which latter Major-General Sir Herbert Stewart was mortally wounded, upon whose death Captain Rhodes was appointed aide-de-camp to Major-General the Honourable T. C. Dormer, commanding a brigade of infantry at Cairo.

In accordance with instructions from the Horse Guards, 7th of September, 1885, 42 undersized horses of the regiment were transferred as under :—20 to 10th Hussars, 15 to 18th Hussars, 7 to 21st Hussars. In the month following 46 more were transferred to the 18th Hussars.

On the 1st of October a troop under Major Rhodes proceeded to Navan, there to be stationed, and on the 9th a troop under Lieutenant Sclater-Booth left Belfast for Dundalk.

On the 8th of October A troop, under Captain Greatrex, proceeded from Dundalk to relieve C troop.

On the 13th of November, Captain Greatrex and Lieutenant Browne, in charge of a detachment of 70 non-commissioned officers and men, proceeded *viâ* Greenmore and Holyhead *en route* to Aldershot. On arrival in London the party divided, one portion proceeding to Aldershot direct, the other to Shorncliffe, where, having received 54 horses from the Queen's Bays, they continued their journey to Aldershot, there to await the arrival of the regiment from Ireland.

On the 14th of November, the 54 horses in question were by Horse Guards authority of the 13th of October, 1885, transferred from the 2nd Dragoon Guards to the Royal Dragoons, to replace those previously transferred to Hussar regiments.

On the 23rd of February, 1886, No. 2067 Private J. Steele received from Her Majesty the Queen at Windsor the medal for distinguished conduct in the field in assisting Lieutenant Burn Murdoch to carry in Lieutenant Guthrie, who had been wounded on the 17th of January previous at the affair of Abu Klea.

On the 29th of April, A and F troops commanded by Captain O'Shaughnessy left Belfast for Dundalk, where, being joined by H troop, the whole under Lieutenant-Colonel A. Maclean proceeded *viâ* Kingstown and Birkenhead, *en route* to Aldershot, arriving there on the 20th of May. C, E, and G troops under Major Maclaren, leaving Dundalk on the 4th of May, embarked at Kingstown on the 6th, and continued *viâ* Birkenhead to Aldershot, arriving there on the 21st. The head-quarters of the Royal Regiment of Dragoons, under Colonel F. Russell, having left Dundalk on the 6th of May, proceeded to Dublin, where, being joined by D troop from Navan, the whole embarked at Kingstown on the 8th, and on landing at Birkenhead the regiment was there detained on the duty of escort to Her Majesty the Queen during her visit to Liverpool, after which they continued to Aldershot, arriving on the 26th of May. The young and sick horses, in charge of Lieutenant-Colonel Rhodes, leaving Dundalk on the 7th and 8th of May, embarked at Kingstown on the 12th, and proceeding *viâ* Portsmouth arrived at Aldershot on the 20th of May.

Before leaving Dundalk Major-General Montgomery Moore had inspected the Royal Dragoons on the 21st of April, who upon parade expressed himself in the highest terms of the good behaviour under most trying circumstances of the non-commissioned officers and men of the regiment, and of his deep regret that they were so soon to leave the country.

On the 2nd of July, 1886, the regiment was present at the review, held by Her Majesty Queen Victoria, of the troops of the Aldershot Division, commanded by Lieutenant-General Sir Archibald Alison, Bart., K.C.B.

On the 3rd of August, the Field-Marshal Commanding-in-Chief H.R.H. the Duke of Cambridge, K.G., inspected the whole of the cavalry and artillery of the Aldershot Division. His Royal Highness lunched with the officers of the Royals at their mess, and was pleased to express in the most flattering terms his high opinion of the good conduct, smart appearance, and excellent physique of the regiment.

On the 1st and 2nd of October, the Regiment was inspected by Major-General Sir D. C. Drury Lowe, K.C.B., Inspector-General of Cavalry in Great Britain, with which satisfactory circumstance may be fitly concluded the first portion of the annals of the Royal Regiment of Dragoons, to be continued, it is hoped, by abler hands.

SUCCESSION OF COLONELS

OF

THE ROYAL REGIMENT OF DRAGOONS.

JOHN, LORD CHURCHILL,

Appointed 19*th of November*, 1683.

At its formation, the Royal Regiment of Dragoons had the honour of being commanded by one of the most distinguished officers Great Britain has ever produced, a general who equally acquired celebrity in the field and in the cabinet, who never fought a battle that he did not win, nor besieged a town that he did not capture.

John Churchill, eldest surviving son of Sir Winston Churchill, Bart., was born on the 24th of June, 1650. At sixteen years of age he was page of honour to H.R.H. the Duke of York, who procured him an ensign's commission in the 1st Foot Guards, but he soon gave up the pleasures of the Court in order to acquire a practical knowledge of his profession at Tangier in Africa, where he served as a volunteer against the Moors, and gave presage of those bright qualities for which he subsequently became so conspicuous. On the breaking out of the Dutch war in 1672, he was appointed captain in the Duke of Monmouth's regiment of foot, in the service of the King of France, with which corps he served in the Netherlands, where he signalised himself by a regular attention to his duties, and by volunteering his services on occasions of difficulty or danger, and especially at the siege of Maestricht in June, 1673, where, as a volunteer with the Life Guards, he was wounded on the 24th of the month. He afterwards served with the French army on the Rhine, where he attracted the particular attention of Marshal Turenne, and in 1674 he was appointed colonel of one of the English regiments in the service of Louis XIV., in succession to the Earl of Peterborough. In 1678 his regiment being recalled from France, Colonel Churchill was appointed to the command of a Brigade of Infantry in Flanders, but the Peace of Nimeguen taking place the same year, he returned to England and his regiment was disbanded.

On the 21st of December, 1682, Colonel Churchill was created Baron Churchill of Aymouth in the Peerage of Scotland, and King Charles II. at this period resolving to increase the regular army, his lordship was commissioned to raise a troop of dragoons, which, being incorporated with one raised by Viscount Cornbury, and with 4 troops of Tangier Horse, the whole were constituted a regiment with the distinguished title of "The Royal Regiment of Dragoons," of which Lord Churchill was appointed Colonel on the 19th of November, 1683.

On the 14th of May, 1685, his lordship was created by James II. Baron Churchill of Sandridge in the Peerage of the United Kingdom, and on the same day was promoted to the rank of Brigadier-General. His lordship, on the breaking out of the rebellion of the Duke of Monmouth, was employed in the West of England in command of a body of cavalry, and at the affair of Sedgemoor, on the 6th of July, 1685, he was second in command of His Majesty's troops. His conduct throughout these disturbances was rewarded by promotion to the rank of major-general, and by the colonelcy of the 3rd troop of Life Guards, being succeeded in that of the Royal Dragoons by Viscount Cornbury.

Edward Viscount Cornbury.

Appointed 1st *of August*, 1685.

Edward Hyde, Viscount Cornbury, son of the second Earl of Clarendon, was appointed to the lieutenant-colonelcy of the "Royal Regiment of Dragoons" when that corps was first constituted, in November, 1683, and having distinguished himself at Sedgemoor on the 6th of July, 1685, he was promoted in succession to Major-General Lord Churchill to the colonelcy of the regiment. The circumstances of his lordship's removal from that post are related at pages 26 to 29 of the history of the corps.

Robert Clifford,

Appointed 24th of November, 1688.

Major Robert Clifford, being firmly attached to the Roman Catholic interests, on the 11th of November, 1688, by his exertions, as will be seen at page 27 of the records of the regiment, succeeded in recovering its services for King James II., who conferred upon him the colonelcy of the corps, of which honour, however, he was deprived by H.R.H., the Prince of Orange, on the 30th of December ensuing, His Royal Highness at the same time re-appointing the Viscount Cornbury, who for political reasons was again suspended a few months afterwards, when Lieutenant-Colonel Anthony Hayford was appointed to the colonelcy on the 1st of July, 1689.

Anthony Hayford,

Appointed 1*st of July*, 1689.

This officer served in the Life Guards as a private gentleman, and afterwards in the Duke of Monmouth's regiment of horse, in the reign of Charles II. In 1684, he was appointed lieutenant in the Horse Grenadier Guards, and in 1687, we find him Lieutenant-Colonel of the Royal Dragoons. He joined the Prince of Orange in 1688, and succeeding Viscount Cornbury in the colonelcy of the regiment in 1689, he served in Scotland and Ireland.

EDWARD MATTHEWS,

Appointed in June, 1690,

served as a volunteer at Tangier, and in 1690 in Ireland, where, having distinguished himself on several occasions as Lieutenant-Colonel of "Leveson's" Dragoons, now the 3rd K.O. Hussars, he was promoted to the colonelcy of the Royal Regiment of Dragoons in the month of June of that year. Subsequently in the years 1694, '95, and '96, he commanded different brigades of dragoons under King William III. in Flanders, and died on the 23rd of May, 1697.

THOMAS, LORD RABY.

Appointed 30*th of May,* 1697.

Thomas Wentworth, son of Sir William Wentworth, Bart., was appointed to a cornetcy in the 4th English Horse, now 3rd Prince of Wales's Dragoon Guards, on the 30th of December, 1688, and in the ensuing summer he served with that regiment against the rebel Highlanders in Scotland. In 1692 he served in Flanders, where at the battle of Steenkirk, on the 3rd of August that year, he distinguished himself with the advanced guard, his squadron being exposed to a heavy cannonade, and bringing out of action but 80 men out of 150. His gallantry upon this occasion having been specially reported to his sovereign, he was appointed aide-de-camp to King William III., in which capacity he was present at the battle of Landen, on the 19th of July,

1693, where his conduct obtained the approbation of His Majesty, who promoted him to the commission of cornet and major in the 1st troop, now 1st regiment of Life Guards.

Major Wentworth served with the Life Guards in the subsequent campaigns in the Netherlands, and rose to the rank of lieutenant and lieutenant-colonel. On the decease of William, Earl of Strafford, he succeeded to the Barony of Raby. On the 30th of May, 1697, his lordship was appointed to the colonelcy of the Royal Dragoons, and attended the Earl of Portland in his interview with Marshal Boufflers, which preceded the conclusion of the Peace of Ryswick in the same year.

In 1698, Lord Raby accompanied King William to Holland, where, in a single-handed encounter with a wild boar, he had a narrow escape for his life, being saved by the timely arrival of 2 huntsmen, sent by the king, who was of the party.

In the first year of the reign of Queen Anne, his lordship served with his regiment on the Continent, and in 1703 he was promoted to the rank of brigadier-general. In the spring of the same year, he was appointed Envoy-Extraordinary to the King of Prussia, and subsequently Ambassador-Extraordinary to the same Court, and on the 1st of January, 1705, he was advanced to the rank of major-general.

Lord Raby served in the army under the Duke of Marlborough, in the brilliant campaign of 1706, in Flanders, and on the 1st of May, 1707, he became lieutenant-general. In 1711, he was sworn of the Privy Council, and was appointed Ambassador-Extraordinary to the States-General of Holland. In September, the same year, he was advanced to the dignity of Earl of Strafford.

The Earl took an active part in negotiating the Treaty of Utrecht on the 1st of April, 1713; but after the accession of George I. on the 11th of August, 1714, he was removed from his public employments, including the colonelcy of the Royal Regiment of Dragoons, conferred on the 13th of June, 1715, upon Lieutenant-General Lord Cobham.

The Earl of Strafford died on the 18th of November, 1739.

RICHARD LORD COBHAM.

Appointed 13*th of June*, 1715.

Sir Richard Temple, of Stowe, Bart., served under King William III. in the Netherlands, and on the breaking out of the War of the Spanish Succession, in 1703, he was promoted to the colonelcy of a newly-raised regiment of foot, which was disbanded at the Peace of Utrecht. He served under the great Duke of Marlborough, and was conspicuous for a noble bearing, and an utter contempt of danger, which he exhibited in a signal manner at the sieges of Venloo, and Karemonah, at the battle of Oudenarde, and at the capture of the city and fortress of Lisle, in 1708.

In January, 1709, he was promoted to the rank of

major-general; and his conduct at the siege of Tournay, the sanguinary battle of Malplaquet, on the 12th of September, and the siege of Mons, was rewarded in the following year by the rank of lieutenant-general and the colonelcy of the 4th "Queen's Own" Dragoons. In 1711, he served under the Duke of Marlborough at the forcing of the French lines at Arleux, an operation performed without the loss of a single British soldier, and at the subsequent capture of the strong fortress of Bouchain in the French department du Nord.

Upon the change of the ministry, and the adoption of a new system of foreign politics, at the end of the reign of Queen Anne, Sir Richard Temple, whose attachment to the Protestant succession was well known, was removed from his regiment; but upon the accession of King George I. on the 11th of August, 1714, on the 13th of June the year following he was appointed colonel of the Royal Regiment of Dragoons. In 1717, the government of Windsor Castle was conferred upon him; and on the 23rd of May, 1718, he was raised to the peerage as Viscount and Baron Cobham. In April, 1721, his lordship was removed to the "King's" Horse, the present 1st or King's Dragoon Guards.

Sir Charles Hotham, Bart.

Appointed 10th of April, 1721.

Charles Hotham, eldest son of the Rev. Charles Hotham, Rector of Wigan, succeeded to the baronetcy on the death of his uncle in 1691. He served with distinction in the wars of King William III., as also under the Duke of Marlborough in the reign of Queen Anne, and in 1703 obtained the colonelcy of a regiment of foot, with which he went to Spain in 1706, and was in garrison at Alicant when the unfortunate battle of Almanza was fought, on the 28th of April, 1707. Sir Charles served with reputation, but his regiment, having suffered severely, was disbanded in Calabria in 1708.

In 1710 he was advanced to the rank of brigadier-general, and shortly after the accession of George I. on the 11th of August, 1714, he was commissioned to raise a regiment of foot, which, after the suppression of the Earl of Mar's rebellion in Scotland in 1715, was sent to Ireland, and there reduced the year following, when Sir Charles Hotham was appointed to the colonelcy of a newly-raised regiment of dragoons, which, however, was reduced in November, 1718. On the 7th of July, 1719, the colonelcy of the 36th Foot was conferred upon him; whence, in December, 1720, he was removed to the 8th Foot; and on the 10th of April, 1721, to the Royal Dragoons.

Sir Charles Hotham died on the 8th of January, 1723.

Humphrey Gore.

Appointed 12*th of January*, 1723.

This officer entered the army as ensign in 1689, and saw much service in the campaigns of King William III.

on the Continent. On the 1st of February he was appointed colonel of a newly-raised regiment of foot, with which in 1709 he went to Spain, and was promoted on the 1st of January, 1710, to the rank of brigadier-general. He was present at the battles of Almanza on the 27th of July, and Saragossa the 25th of August, but on the 6th of December following, he was taken prisoner in the unfortunate affair of Brihuega.

At the peace of Utrecht, 11th of April, 1714, "Gore's" regiment of foot was disbanded; but in July, 1715, being a firm adherent to the Protestant succession at a time when Jacobite principles became very prevalent, Brigadier-General Gore was commissioned by His Majesty King George I. to raise a regiment of dragoons, which has become the present 10th Royal Hussars.

On the 12th of January, 1723, he was transferred to the Royal Regiment of Dragoons. He became major-general on the 6th of March, 1727, lieutenant-general the 27th of October, 1735, and died on the 18th of August, 1739.

Charles, Duke of Marlborough, K.G.

Appointed 1st of September, 1739.

Charles Spencer, fourth Earl of Sunderland, succeeded to the Dukedom of Marlborough in 1733, and five years later he was appointed to the colonelcy of the 38th Foot. In 1739 he was removed to the Royal Dragoons, and the year following to the 2nd Troop of Life Guards.

Henry Hawley.

Appointed 12th of May, 1740.

This remarkable officer served the Crown in four successive reigns, and held a commission in the army for a period of sixty-five years. His first appointment is dated 10th of April, 1694, and having signalised himself in the wars of Queen Anne, he obtained the rank of colonel by brevet on the 16th of October, 1712. In 1715, during the rebellion in Scotland, he was wounded in the affair at Dunblane on the 13th of November. On the 19th of March, 1717, he was promoted from the lieutenant-colonelcy of the 4th Queen's Own Dragoons to the colonelcy of the 33rd Foot, from which, on the 7th of July, 1730, he was transferred to that of the 13th Dragoons. In 1735 he was advanced to the rank of brigadier-general; in 1739 to that of major-general; and the year following he was appointed to the colonelcy of the Royal Regiment of Dragoons.

In 1742 Major-General Hawley proceeded with the army commanded by Field-Marshal the Earl of Stair to Flanders. He was promoted the following year to the rank of lieutenant-general, and was present at the battles of Dettingen and Fontenoy.

In 1745 the Lieutenant-General was again in Scotland, during the rising in that country in favour of Prince Charles Edward, and in an affair with the rebels at Falkirk on the 10th of May, 1746, the troops under his command were defeated with considerable loss. The

General served afterwards on the staff in Ireland, and for many years was Governor of Portsmouth. He died on the 24th of March, 1759.

The Honourable Henry Seymour Conway.

Appointed 5*th of April*, 1759.

The Honourable Henry Seymour Conway, second son of Lord Conway, and brother of Francis 1st Earl of Hertford, was appointed lieutenant in the 1st Foot Guards in 1737, captain and lieutenant-colonel in 1741, and in 1746 he was appointed aide-de-camp to H.R.H. the Duke of Cumberland, and promoted to the colonelcy of the 59th, now 48th Foot. In 1749 he was removed to the 34th Foot, to the 13th Dragoons in 1751, and to the 4th Irish Horse, now the 7th P. R. Dragoon Guards, in 1754. He was promoted to the rank of major-general in 1756, and in 1759 to that of lieutenant-general, in which year also he was transferred to the colonelcy of the Royal Dragoons.

Lieutenant-General Seymour Conway commanded a division of the allied army in Germany under H.S.H. the Duke of Brunswick in 1761, and during the absence of the Marquis of Granby the British troops in that country were placed under his orders.

In 1747 he held the government of Jersey. He was one of the Grooms of the Bedchamber to His Majesty George III., a Privy Councillor, and a member of Parliament; but in 1764, having voted against ministers on the great question of military warrants, he resigned his civil appointments and military commands, including the colonelcy of the Royal Regiment of Dragoons, which, on the 9th of May that year, was conferred on

HENRY, EARL OF PEMBROKE,

the 10th earl, who, entering the army in 1752, in 1754 obtained a troop in the 1st or King's Dragoon Guards. In 1756 he was promoted captain and lieutenant-colonel in 1st Foot Guards, and on the 8th of May, 1758, his lordship was appointed aide-de-camp to King George II., with the rank of colonel.

In the following year, being appointed lieutenant-colonel of the 15th Light Dragoons, the present 15th King's Hussars, he accompanied his regiment to Germany, and served with distinction under the Marquis of Granby during the remainder of the Seven Years' War.

In 1761 his lordship attained the rank of major-general, and on the 9th of May, 1764, his Majesty George II. bestowed upon him the colonelcy of the Royal Regiment of Dragoons. On the 30th of April, 1770, he was promoted to the rank of lieutenant-general, and to that of general in November, 1782. His lordship for many years held the government of Portsmouth, and died on the 26th of January, 1794. The Earl of Pembroke was the author of an excellent work on horsemanship.

PHILIP GOLDSWORTHY.

Appointed 28th of January, 1794.

This officer was many years in the Royal Dragoons, with which corps he served in Germany during the Seven Years' War. On the 18th of April, 1779, he was promoted to the lieutenant-colonelcy of the regiment, obtained the rank of major-general on the 20th of December, 1793, and in the month following succeeded the Earl of Pembroke in the colonelcy of the regiment. On the 26th of June, 1799, he was promoted to the rank of lieutenant-general, and died in 1801.

Lieutenant-General Philip Goldsworthy was an equerry and clerk marshal to King George III., and M.P. for Wilton.

THOMAS GARTH.

Appointed 7th of January, 1801.

Appointed to a cornetcy in the Royal Regiment of Dragoons on the 12th of April, 1762, he served in the campaign of that year with his regiment in Germany. Promoted lieutenant in 1765, and captain in 1778, he exchanged in 1779 to the 20th Light Dragoons, with which corps he served many years in the West Indies. In 1792 he was promoted to a majority in the 2nd Dragoon Guards, "Queen's Bays," and in 1794 to a lieutenant-colonelcy in the Royal Dragoons.

Having served in Flanders with the army under H.R.H. the Duke of York, he was rewarded by the colonelcy of the "Sussex Fencibles," from which he was removed to the 22nd Light Dragoons. Promoted to the rank of major-general in 1798, he was transferred to the Royal Dragoons in 1801. He obtained the rank of lieutenant-general in 1805, and that of general in 1814, and died 18th of November, 1829.

General Garth was for many years an equerry to His Majesty George III.

Lord Robert Edward Henry Somerset, G.C.B.

Appointed 23*rd of November*, 1829.

The third son of Henry fifth Duke of Beaufort, his lordship was appointed on the 11th of February, 1793, to a cornetcy in the 10th Light Dragoons, commanded by H.R.H. the Prince of Wales, afterwards King George IV.

He became lieutenant in December following, and captain in August, 1794. In 1798 he held the appointment of aide-de-camp to the Lieutenant-General commanding the Severn District, and stationed at Bristol, but in the year following he accompanied H.R.H. the Duke of York to Holland, where his Royal Highness held the command of the British army, and where, in the position of aide-de-camp, his lordship made himself conspicuous by his zeal and activity, particularly in the action at Bergen on the 19th of September, 1799, at Egmont-on-Zee on the 2nd of October, and in the subsequent military operations in the United Provinces. His lordship this same year was returned to Parliament for the Monmouth Boroughs, which he continued to represent until 1802, and upon his return to England he was promoted to a majority in the 12th Light Dragoons, with which corps he was serving in Portugal, when in November, 1800, he was transferred to the 28th Light Dragoons, and on the 25th of December ensuing he was promoted to the lieutenant-colonelcy of the 5th Foot. On the 28th of September, 1801, he was transferred to the 4th Queen's Own Dragoons, which regiment he commanded for many years with great distinction.

In 1803 Lord Edward Somerset was elected to Parliament for the county of Gloucester, which he continued to represent until 1829.

Early in 1809 his lordship accompanied the eight troops of the 4th Queen's Own Dragoons to Portugal, there to join the army under Viscount Wellington, and was engaged at the battle of Talavera on the 28th of July that year.

In July, 1810, he was appointed aide-de-camp to His Majesty George III., with the rank of colonel in the army. His eminent exertions at the cavalry affair at Usagre on the 28th of May, 1811, and his conduct in the brilliant charge of the heavy dragoons at Salamanca on the 22nd of July, 1812, is immortalised in the history of the campaigns in the Peninsula.

Promoted to the rank of major-general in June, 1813, his lordship, upon this separation from the regiment with which he had so long been conspicuously identified, was presented by the officers of the 4th Queen's Own Dragoons with a valuable mark of their esteem and regard in the shape of a sword of honour.

Placed in command of a Hussar brigade of the 7th, 10th, and 15th regiments, Lord Edward Somerset continued with the army in the Peninsula until the close of the war in 1819, and was present at the battle of Vittoria during the operations in the Pyrenees, and at the battles of Orthes and Toulouse.

On the 26th of July his lordship had the honour of receiving in his place in the House of Commons the thanks of the House for his eminent services in the Peninsular War.

In the spring of 1815, on the return of Napoleon to France and the consequent renewal of hostilities, Major-General Lord Edward Somerset was placed upon the staff of the army in Flanders under Field-Marshal the Duke of Wellington, and appointed to the command of the first brigade of cavalry, consisting of the the 1st and 2nd Life Guards, the Royal Horse Guards, and the 1st or King's Dragoon Guards. The conduct of that brigade on the transcendent day of Waterloo is too well known to need any recapitulation, and at the head of those splendid regiments Lord Edward Somerset more than equalled his previous reputation. For his services on that glorious occasion his lordship, on the 29th of April, had again the honour of acknowledging in the House of Commons the thanks of that House.

On the 18th of January, 1818, the Major-General was appointed to the colonelcy of the 21st Light Dragoons, and remained in command of a brigade of cavalry with the army of occupation in France until, in the course of the same year, the troops returned to England.

His lordship was now appointed Inspector-General of Cavalry, which position he held until his promotion to the rank of lieutenant-general, in May, 1825.

On the 9th of September, 1822, he was removed to the colonelcy of the 17th Lancers, and on the 23rd of November, 1829, to that of the Royal Regiment of Dragoons.

During the years 1829 and 1830 Lord Edward held the appointment of Lieutenant-General of the Ordnance, and for a short time in 1835 that of Surveyor-General of that department. In 1836 he was transferred to the 4th Queen's Own Dragoons.

Lord H. Edward Somerset received the gold cross with one clasp for services in command of the 4th Dragoons at the battles of Talavera and Salamanca, and as Major-General commanding a brigade at Vittoria, Orthes, and Toulouse, and the medal for Waterloo. He was a Grand Cross of the Order of the Bath, a Knight of the Tower and Sword of Portugal, of Maria Theresa of Austria, and of St. Vladimir of Russia.

THE HONOURABLE SIR FREDERICK CAVENDISH PONSONBY, K.C.B., G.C.M.G., K.C.H.

The second son of Frederick, third Earl of Bessborough, he was appointed cornet in the 10th Light Dragoons in 1800, in which corps obtaining a troop in 1803, he exchanged in 1806 to the 60th Foot. In 1807 he was promoted to a majority in the 23rd Light Dragoons, with which regiment he was present at the battle of Talavera in 1809, and obtained the lieutenant-colonelcy in the following year.

In 1811, while serving on the staff under Lieutenant-General Graham at Cadiz, at the battle of Barossa, in March of that year, he attacked, with a squadron of German Dragoons, the French cavalry covering the retreat, overthrew them, took two guns, and even attempted to cut down the battalions of General Rousseau. On the 11th of June the same year he was appointed lieutenant-colonel of the 12th Light Dragoons, with which corps he served under Wellington, and with particular distinction at the brilliant cavalry affair at Llerena, in April, 1812. At the battle of Salamanca, in July following, in charging the French infantry his sword was broken, and his horse severely wounded by the bayonets. In the retreat of the army from Burgos he was wounded on the 13th of October, having repeatedly shown remarkable instances of judgment and decision in outpost duties.

At the battle of Vittoria, 20th of June, 1813, he again signalised himself, as well as in the subsequent operations at Tolosa, St. Sebastian, and the Nive, and on the king's birthday in 1814 he was promoted to the rank of colonel in the army.

In the memorable campaign of 1815 in Flanders, Colonel Ponsonby commanded the 12th Light Dragoons at Waterloo, in which great conflict he showed signal intrepidity, and was very severely wounded, being disabled by sabre cuts in both arms, knocked off his horse by a blow on the head, run through the body by a lance, ridden over by two squadrons of cavalry, and twice plundered as he lay defenceless on the ground.

In January, 1824, Colonel Ponsonby was appointed Inspecting Field Officer in the Ionian Islands, being promoted to the rank of brigadier-general on the staff of these islands on the 4th of March ensuing. In June, 1825, he became major-general, and was removed to the staff at Malta, where he commanded the troops until May, 1835, in which year the colonelcy of the 86th Foot was conferred upon him, whence he was transferred to that of the Royal Regiment of Dragoons in March, 1836, and died on the 10th of January, 1837, having been in every sense of the word a conspicuous ornament to his profession.

Sir Frederick Ponsonby had been rewarded for his services by the following distinctions:—Knight Commander of the Order of the Bath, Knight Grand Cross of the Order of Saint Michael and Saint George, Knight

Commander of the Royal Hanoverian Guelphic Order, a gold cross for battles in the Peninsula, the medal for Waterloo, Knight of the Tower and Sword of Portugal, and Knight of Maria Theresa of Austria.

The Right Honourable Sir Richard Hussey Vivian, G.C.B., G.C.

Appointed 20th of January, 1837.

Entering the army on the 30th of November, 1793, as ensign in the 20th Foot, he was promoted to captain in the 28th Foot on the 9th of May, 1794, and on the 3rd of August, 1798, he was transferred to the 7th "Queen's Own" Light Dragoons. He became major and brevet lieutenant-colonel on the 9th of March, 1803, and he was appointed to the command of the regiment on the 1st of December, 1804.

In the month of October, 1808, Lieutenant-Colonel Vivian embarked at Portsmouth in command of the 7th Q.O. Hussars for Portugal, and on the 21st of December the regiment, with the 10th and 15th Hussars, under Lieutenant-General Lord Paget, was engaged with the French cavalry at Benevente, when, in a sharp encounter of about 20 minutes, 20 of the enemy were killed, and 13 officers and 104 men were taken prisoners.

Embarking at Corunna, after the battle there of the 10th of January, 1809, the 7th Hussars returned to England.

On the 20th of February, 1812, Lieutenant-Colonel

Vivian was promoted to the rank of colonel in the army, and was appointed in the same year an equerry to H.R.H. the Prince Regent, afterwards King George IV.

On the 10th of August, 1813, the 7th Hussars, under Colonel Vivian, embarked a second time for the Peninsula, when 8 troops landing at Bilbao on the 1st of September were joined in the following month by two other troops from England.

Colonel Vivian was now appointed to the command of a brigade, consisting of the 18th Hussars and 1st Hussars K.G.L., which he held during the remainder of the campaign of that year, and at the battle of Orthez on the 27th of February, 1814.

On the 8th of April, 1814, at the brilliant affair at Croix d'Orade, in which the 18th Hussars, led by Major Hughes, were so distinguished, Colonel Vivian was severely wounded by a carbine shot. He was promoted to the rank of major-general on the 4th of June following, receiving at the same time the honour of Knight Commander of the Bath.

Upon the general renewal of hostilities in 1815, Major-General Sir R. Hussey Vivian, K.C.B., on the 10th of March was appointed to command a brigade, 10th and 15th Hussars and 1st Hussars K.G.L. in the army in the Netherlands under Field-Marshal the Duke of Wellington. At the great battle of Waterloo, Sir Hussey Vivian, with his arm in a sling and still suffering from the wound received at Croix d'Orade, more than a year previous, was attacked by a French cuirassier when, dropping his reins and drawing his sword with his left arm, he wounded the man in the throat, and his orderly, a hussar of the German Legion, coming up at the moment, cut the cuirassier down. The charge of Vivian's Brigade on this glorious day forms one of the leading episodes of the battle.

On the 25th of June Major-General Sir R. Hussey Vivian, K.C.B., and the officers and men under his orders, received the honour of a vote of thanks from both Houses of Parliament for their gallant conduct at Waterloo.

In 1820 the Major-General was returned to Parliament for Truro, which borough he represented until 1826.

On the 10th of September, 1821, the 18th Hussars were disbanded, upon which occasion a very interesting memorial was presented by the regiment to Major-General Sir R. Hussey Vivian, K.C.B., being a silver trumpet with banner, purchased by desire of the soldiers of the 18th with part of the prize money accruing from the horses of the enemy, captured by the brigade under the command of Major-General Sir R. Hussey Vivian, K.C.B., at the battle of Waterloo, and with the following inscription :—

" On the 10th of September, 1821, the day on which the 18th Hussars was disbanded, this trumpet was pre-

S 2

sented to Major-General Sir Hussey Vivian, K.C.B. Having commanded them upon many glorious occasions, they offer to him this memorial of the last victory in which it was their fortune to be led by him, as an assurance that while he gained their admiration as a soldier, he secured their lasting and unfeigned esteem as a friend, and in the hope of living in his recollection and estimation when they shall have ceased to exist as a corps."

On the 10th of September, 1880, the 59th anniversary of the disbandment of the regiment, this trumpet was generously entrusted by Charles Crespigny, 2nd Baron Vivian, to the care of the 18th Hussars, "believing that in this record of glorious deeds, the memory of his father, who led the regiment to victory on many occasions, will be cherished in the corps whose admiration he secured."

The Major-General was appointed on the 1st of February, 1825, Inspector-General of Cavalry, and held that post until the 20th July, 1830.

In 1826 he was returned to Parliament for Windsor, which Borough he represented till 1831.

On the 22nd of January, 1827, he became lieutenant-general in the army, and the day following the colonelcy was conferred upon him of the 12th Royal Lancers. On the 29th of January, 1828, he was created a baronet. He was appointed to the command of the forces in Ireland on the 1st of July, 1831.

In 1832 he was again returned to Parliament for

Truro, until in 1837 he was returned for Cornwall, which county he represented until 1841.

In 1834, upon the occasion of the Duke of Wellington being elected Chancellor of the University of Oxford, the honorary degree of D.C.L. was conferred upon Lieutenant-General Sir R. Hussey Vivian, Bart., K.C.B., &c.

From May, 1835, to September, 1841, he held the office of Master-General of the Ordnance. He was transferred on the 20th of January, 1837, to the colonelcy of the "Royal Regiment of Dragoons," and on the 19th of August, 1841, he was raised to the peerage as Baron Vivian of Glynnand, Truro, Cornwall.

His Lordship died on the 20th of August, 1842.

Lord Vivian was a Privy Councillor, and had received the distinctions of a—

Grand Cross of the Order of the Bath; Grand Cross of the Order of St. Michael and St. George; Grand Cross of the Royal Hanoverian Guelphic Order; Knight of Maria Theresa of Austria and of St. Vladimir of Russia; a gold medal with clasp for the affairs at Benevente and Orthez, and the medal for Waterloo.

His lordship had also granted to him, by the special desire of His Majesty George IV., an augmentation crest for the action of Croix d'Orade.

Sir Arthur B. Clifton, G.C.B., K.C.H., K.St.W., K.M.T.

Appointed 30th of August, 1842.

The third son of Sir Gervase Clifton, Bart., of Clifton Park, Nottinghamshire, where also he himself possessed a handsome estate. He was appointed to a cornetcy in the 3rd Prince of Wales' Dragoon Guards on the 6th of June, 1794, lieutenant on the 7th of August following; captain on the 27th of February, 1799, and major on the 17th of December, 1803.

Accompanying his regiment to Portugal in 1809, Major Clifton was present with it at Talavera on the 28th of July that year; and when, on the 13th of July, 1810, Lieutenant-Colonel Windham of the Royal Dragoons was taken prisoner, Lieutenant-General Sir Stapleton Cotton, Bart., commanding the cavalry in the Peninsula, recommended Major Clifton for the lieutenant-colonelcy of that regiment, which, being confirmed on the 22nd of November ensuing, he commanded it with much distinction to the close of the war in 1814.

Throughout those campaigns the name of Lieutenant-Colonel Clifton will be found frequently and honourably mentioned. He was present at Busaco on the 27th of September, 1810. At Fuentes d'Onoro on the 5th of May, 1811, two squadrons of the Royal Dragoons, commanded by Lieutenant-Colonel Clifton, made a gallant and successful charge upon the enemy's cavalry, took a sergeant and 23 men, and released a party of the Foot Guards who had been made prisoners by the French.

In the affair at Gallegos, on the 6th of June, 1812, and on the 11th of the same month at Llera, the conduct of Lieutenant-Colonel Clifton is again specially signalized; and on the 26th of May, 1813, in the neighbourhood of

Salamanca, the right squadron of the Royals, led by Lieutenant-Colonel Clifton, charged the enemy with brilliant effect, sabred a number of men, and took 143 prisoners, with four tumbrils.

At the battle of Vittoria on the 21st of June following, Lieutenant-Colonel Clifton was in command of a Brigade; and having assisted at the battle of Toulouse on the 10th of April, 1814, the closing scene of the Peninsular war, he served in the memorable campaign of 1815 in Flanders, when at the battle of Waterloo, succeeding to the command of the "Union" Brigade of Cavalry, he marched with it to Paris.

On the 12th of August, 1819, he became colonel in the army, and retiring upon half-pay on the 10th of June, 1829, he was promoted to the rank of major-general on the 22nd of July, 1830.

The colonelcy of the 17th Lancers was conferred upon him on the 24th of August, 1839. He became lieutenant-general on the 23rd of November, 1841; and on the 25th of April, 1842, he was transferred to the 11th Prince Albert's Hussars, from which, on the 13th of August ensuing, he was removed to the Royal Regiment of Dragoons.

He attained the rank of general on the 20th of July, 1854.

It may be noticed that when colonel of the Royal Dragoons, Sir Arthur Clifton had been offered the colonelcy of the 1st or King's Dragoon Guards, at that period stronger than other regiments of cavalry, and worth £400 per annum more in the way of emolument; he declined this favour, preferring to remain with the corps with which he had so long been associated, and in which he had acquired a reputation so well deserved.

Sir Arthur Clifton died on the 7th of March, 1869, at the age of 97.

The General had received the Grand Cross of the order of the Bath; Knight Commander of the Royal Hanoverian Guelphic order; a gold medal with one clasp for Fuentes d'Onoro and Vittoria; silver medal with three clasps for Talavera, Busaco, and Toulouse; medal for Waterloo; Knight of Maria Theresa of Austria; Knight of St. Vladimir of Russia; Knight of St. Anne of Russia; fourth class Wilhelm of the Netherlands.

Charles Philip de Ainslie.

Appointed 8th of March, 1869.

The representative of the ancient family of De Ainslie of Dilphington, Roxburghshire, N.B., of which many members had served in the army, this officer was appointed to a second lieutenancy in the Rifle Brigade on the 10th of April, 1825; and served with the 1st battalion in Halifax, N.S., until promoted on the 28th of January, 1826, to a lieutenancy unattached; and the day following, exchanged to the 4th "Queen's Own" Light Dragoons, the depot of which corps he joined at Maidstone.

From June, 1827, to May, 1828, he served as aide-de-camp to Lieutenant-General Sir Thomas Bradford, K.C.B., commander-in-chief at Bombay, when returning to England, he joined the Cavalry Riding Establishment at St. John's Wood, until being promoted on the 16th of March, 1830, to a company unattached, he exchanged on the 29th of June following to the 1st or Royal Regiment of Dragoons.

From May, 1840, to 1st of April, 1842, Captain de Ainslie served as aide-de-camp to Major-General Lord Greenock, K.C.B., commanding the forces in Scotland, and on the 14th of October that year, he purchased the majority of the Royal Dragoons.

On the 3rd of February, 1843, exchanging to the 14th King's Light Dragoons, he joined that regiment at Bombay; and was promoted to an unattached lieutenant-colonelcy on the 27th of October, 1847.

On the 23rd of February, 1849, Lieutenant-Colonel de Ainslie exchanged to the 7th P. R. Dragoon Guards, and commanded that corps until he retired upon half-pay on the 8th of December, 1854, having been promoted to the rank of colonel by brevet of the 28th of November previous.

On the 25th of August, 1857, Colonel de Ainslie re-changed to full pay in the 14th King's Light Dragoons; and joining that corps in India, he there commanded with the rank of brigadier, the cavalry at Kirkee, and subsequently a brigade at Jhansi in Central India during the Sepoy Mutiny. The 14th Light Dragoons returning to England, he was placed on half-pay as second lieutenant-colonel on the 28th of August, 1860.

Advanced to the rank of major-general on the 7th of July, 1862, he was appointed in May, 1866, to the command of the troops in the Windward and Leeward Islands, with head-quarters at Barbadoes, and upon the consolidation of the two commands, to that of the forces in the West Indies, with head-quarters at Jamaica.

On the 8th of March, 1869, receiving the colonelcy of the Royal Regiment of Dragoons, the major-general resigned the command in the West Indies, and returned to England in May the same year. He became lieutenant-general on the 25th of October, 1871, and general on the 12th of October, 1877.

In the year 1846, General de Ainslie attended at Luneville some French cavalry manœuvres upon a new system, to which he invited attention, and published his own opinions, and which ultimately led to a radical change in the principles of the manœuvres of the British cavalry.

INDEX.

D

F

I

M

O

Q

R

S

U

V

W

THE END.

www.ingramcontent.com/pod-product-compliance
Ingram Content Group UK Ltd.
Pitfield, Milton Keynes, MK11 3LW, UK
UKHW021837270726
14058UKWH00002B/212